MOUNTAINS AND NORTHERN FORESTS

BRITISH COLUMBIA: A NATURAL HISTORY

MOUNTAINS AND NORTHERN FORESTS

Richard Cannings and Sydney Cannings

GREYSTONE BOOKS

DOUGLAS & McINTYRE

VANCOUVER/TORONTO

98 99 00 01 02 5 4 3 2 1

Greystone Books
A division of Douglas & McIntyre Ltd.
1615 Venables Street
Vancouver, British Columbia
V5L 2H1

Canadian Cataloguing in Publication Data
Cannings, Richard J. (Richard James)
Mountains and northern forests

ISBN 1-55054-636-8

1. Natural history—British Columbia, Northern. 2. Mountain ecology—British Columbia. 3. Taiga ecology—British Columbia. I. Cannings, Sydney G. (Sydney Graham), 1954- II. Title.
QH106.2.B7C358 1998 508.711'8 C97-911098-X

Editing by Nancy Flight
Front cover photograph by Richard Cannings
Back cover photograph by Douglas Leighton
Cover and text design by George Vaitkunas
Maps by Eric Leinberger
Printed and bound in Canada by Friesens

The publisher gratefully acknowledges the assistance of the Canada Council and of the British Columbia Ministry of Tourism, Small Business and Culture. The publisher also acknowledges the financial support of the Government of Canada through the Book Publishing Industry Development Program for its publishing activities.

***Page ii:* Winter forest, Glacier National Park, British Columbia**

CONTENTS

PREFACE

The mountains and northern forests of British Columbia offer countless spectacular destinations for the campers, hikers and other outdoor enthusiasts that are fortunate enough to live in or visit this beautiful province. Like the rest of the natural world, the mountains and forests also have many stories to tell—stories of how plants and animals survive the world's deepest snowfalls, how tiny birds migrate from northern British Columbia to Costa Rica fuelled by a few grams of fat and guided by genetic memories, how tiny flies find mates in the vast wilderness. Any adventure in the natural world can recharge human batteries drained by the stresses of modern life, and knowing these stories can enrich the adventure even further.

When we wrote *British Columbia: A Natural History,* we realized that the subject was too vast for a single book. We tried to paint the big picture of how plants and animals flourish in the many different habitats of British Columbia—including rain forests, cold forests, dry forests, tundra, grasslands, wetlands and seacoasts—highlighting that broad canvas with a few specific examples. Many stories were left unwritten, and many that were written had to be edited out for space considerations. We decided then to create a series of smaller books from the larger project. Each book would take a chapter or two of the original book and expand on that theme, adding new sections, photographs and illustrations.

This book is one of the first in that series. Since it combines two rather long chapters, its content is perhaps more similar to that of the original book than the content of other books in the series will be, but nonetheless many new stories, photographs and illustrations have been added. Because it is part of a series, there may be noticeable gaps in this book—for instance, geology and aquatic biology—that will be filled by other books in the series.

Most outdoor enthusiasts want to learn more about the world they walk, hike, bike, canoe, kayak or climb through. Until recently there were no books in British Columbia that filled that need beyond the level of the generic field guide. This series is designed to both entertain and educate nature lovers of all kinds, and we hope that it will inspire them to want to know more.

Pages vi–vii: **This view of Chilko and Tsuniah Lakes typifies most of British Columbia—high alpine ridges, valley lakes and rivers, and broad, forest-covered plateaus.**

ACKNOWLEDGEMENTS

THIS BOOK AND THE OTHERS IN THE SERIES were inspired by many years of conversations along forest trails, around campfires and in university coffee rooms. The naturalist tradition in British Columbia is largely an oral tradition, and many very knowledgeable people have freely shared their stories with us. We could not have begun to write this book without them.

We would first like to thank Trevor Goward for his broad knowledge of some of the more inconspicuous members of the natural world and his very constructive criticism of this manuscript. Other reviewers who read all or part of the manuscript include Margaret Holm, Leah Ramsay, Douglas Leighton, Nancy Baron, Terry Taylor, Bette Cannings, Robert Cannings, Ken Lertzman, Gail Kenner, Tony Sinclair, Judy Myers, Don Reid and David Stirling.

We truly appreciate the work of Donald Gunn, whose line drawings are a major part of this book. Other illustrations were provided by Robert Cannings and W. B. Schofield. Many photographers offered their photographs for this work; we particularly thank Douglas Leighton, Mark Hobson, Steve Cannings and Chris Harris. Other images were provided by Dave Fraser, Robert Cannings, Ian Jones and Graham Osborne.

INTRODUCTION

ONE LATE AUGUST DAY we climbed through freshly fallen snow to the top of Mount Tatlow, a soaring double pyramid at the edge of the Chilcotin Mountains. At the icy, 3066-metre summit we were treated to a fabulous vista of the heart of British Columbia. To the south and west lay an awesome display of rock and ice—the granitic core of the Coast Mountains. To the north and east the rolling, almost flat expanse of the Chilcotin Plateau stretched to the horizon, covered by a dark green cloak of pine and spruce.

This stunning view reflected the landscape that covers most of the province. Although most British Columbians live along the coast or in the southern valleys and like to think of British Columbia as a temperate place, the "banana belt" of Canada, the majority of the province is a combination of mountains and high plateaus. This vast area of British Columbia is relatively isolated from the moderating influence of the Pacific Ocean, and the mountain ranges that run its length thrust masses of rock into the cold, thin air. The Interior mountains and plateaus have long Canadian winters; they are the land of the spruce, the realm of Moose and mosquitoes. When the words *wilderness* and *camping* are mentioned, this is the country Canadians usually think of. Most of British Columbia's large parks are predominantly spruce forests and high mountain meadows, where Common Loons cry from clear lakes, Hermit Thrushes sing their flutelike songs from fragrant, swaying fir trees, and Lynx wait in ambush for Snowshoe Hares.

Pages x–1: A line of Alpine Larches seems to march towards treeline on Mt. Ymir in the Selkirk Mountains.

On a provincial scale, the mountain and northern forests of British Columbia have been divided into a number of ecological zones based on climate and dominant trees. Although there are clear differences among these forest zones, and the treeless alpine tundra is a diverse environment itself, they do share much of their flora and fauna and at least one climatic variable—a long, cold winter.

This book will divide this huge area into two parts—the open mountaintops, with their flower-filled meadows, stunted trees and lichen-covered rocks; and the forests that cover the mountains and plateaus with spruce, fir and pine.

PART ONE
MOUNTAINTOPS

MOUNTAINTOPS are magical places. Their wildness, their isolation from the human world and their magnificent vistas evoke strong emotions in all of us. Every hiker has his or her own magical memories—our favourite is of a glorious end-of-summer evening atop Tonquin Hill, an alpine knoll amid some of the most spectacular scenery in the Rocky Mountains. Leaning into a warm west wind, we surveyed the seemingly endless mountains. From Mount Robson towering above the northern skyline to the headwaters of the Fraser River just below the southern horizon, not one sign of human activity was visible.

A small herd of caribou clattered over the rocks below, and as the sun went down, White-tailed Ptarmigan began to fly into the alpine meadow behind us, cackling to each other as they went to roost.

From most places in British Columbia, even from downtown Vancouver, you can see wave upon wave of mountains. But most British Columbians spend little time on mountain peaks, admiring them from below instead. In more ways than one, however, you get an entirely different view of the world from a mountaintop. We normally see British Columbia as a series of valleys framed by postcard peaks, but if you stand atop even a small mountain in the Rockies, Coast Mountains, Cassiars, Selkirks or Purcells, you will always be impressed by the sea of rugged mountains around you and by how small the lush valleys really are.

Entirely different too is the life at your feet, where cushions of Moss Campion huddle behind rocks and fat bumblebees tack against the wind to the next flower. At lower elevations, competition among organisms is rampant—trees battle among themselves to get more and

Bumblebees on Rosebay Fireweed

Hot Bees, Cold Bees

A cold mountain breeze can discourage mosquitoes and blackflies by cooling their bodies below the minimum temperature for their flight muscles to work, but it doesn't usually faze the big bumblebees. They can shiver until their wing muscles are warm enough to work and then fly to a clump of flowers to refuel. At flowers such as goldenrods, which have many small flowers, each with a small nectar reward, the bees let their bodies cool again and walk around the flower clumps, slowly tanking up. On fireweed and other nectar-rich flowers, however, the bumblebees can afford to keep their wing muscles fired up, ready to fly to the next patch of flowers.

more sunlight and more nutrients from the soil; barnacles and mussels fight on intertidal rocks just to find a place to settle down. Above the timberline, however, most of the competition is between organisms and the harsh environment—plants struggle to grow a few millimetres each summer and maybe even set a few seeds; marmots slip into a deep winter sleep each September, hoping they have put on enough fat to get them through an eight-month fast until the first spring shoots appear in late May.

TIMBERLINE

The natural world is full of boundaries, narrow zones where changes in environmental forces such as moisture or temperature produce a radical shift in the flora and fauna. One of these boundaries, obvious to even the most casual observer, is timberline, the zone between thick forests and barren tundra. Although at close range it can appear to be a relatively wide zone where forests gradually give way to larger and larger meadows and trees become progressively smaller, from a distance timberline is usually remarkably abrupt. The timberline zone can be divided into narrower belts by defining

> . . . BUT THE SOUL OF THE COUNTRY IS OUT PAST THE TIMBERLINE.
>
> —Murray McLauchlan, "Out Past the Timberline"

three boundaries associated with it: the forest line, where continuous forest is invaded by subalpine meadows; treeline, the upper limit of erect trees; and the krummholz line, the upper limit of stunted mats of conifers (Figure 1). Ecologists refer to the region above the treeline as the alpine zone; the forests, woodlands and meadows immediately below the treeline are generally termed subalpine.

What are the forces that produce such a sudden change in the habitat? Cold and snow. It is usually not frigid winter temperatures that govern the elevation of treeline, however. Conifers can withstand temperatures down to -60°C or lower, and winters at treeline in British Columbia are much milder than that, at least in the south. The forces of cold and snow act together in different combinations throughout British Columbia to create three different kinds of timberline: coastal, southern Interior and northern Interior.

FIGURE 1

Generalized mountaintop in southern British Columbia showing the relationship between the forest line, treeline and krummholz line. Note that these occur at lower elevations on the shadier north side of the mountain. The zones defined by these three boundaries form the timberline.

Coastal Timberlines

On the moist west slope of the Coast Mountains, where winters consist of a series of Pacific storms, it is the

sheer amount of snow that governs the upper limit of tree growth. There, that upper limit is usually around 1650 metres, the lowest treeline in southern British Columbia. The winters are rather short and mild in the coastal mountains, with mean January minimum temperatures of about -5°C and frost-free periods of over one hundred days. But this is the snowiest place on earth, with annual snowfalls of more than 15 metres and maximum accumulations of 5 metres or more. The heaviest winter snowfall ever recorded—27.4 metres—occurred in 1971–72 on Mount Rainier, Washington. Similar snowfalls undoubtedly occur in the British Columbia Coast Mountains, but there are no permanent weather stations to measure them. A weather station that operated for six years at 1625 metres elevation above Kemano on the central coast recorded an average annual snowfall of 19.7 metres, more than the average at Mount Rainier.

At treeline, conifer branches can only thrive under the winter snowpack, forming dwarfed shrubs called krummholz—German for crooked wood.

These huge snowpacks take so long to melt that the growing season is too short for tree seedlings to germinate. It is often late July or even August before timberline plants emerge from the snow in coastal mountains, especially in protected bowls where heavy, cold air pools on spring and summer nights, slowing melting even further.

On coastal mountains exposed to the full force of Pacific storms, treeline is even lower, since winds of up to 300 kilometres per hour blow any branches away. In

the mountains of the Queen Charlotte Islands, for example, alpine habitats are encountered as low as 1000 metres elevation.

Southern Interior Timberlines

Snowfall in most of the Interior mountains is much less than on the coast, but winter temperatures are still not harsh enough to be the critical factor shaping treeline. A quick look at the stunted patches of krummholz, with their spreading skirts protected under the snow and every attempt at growth above the snow killed back, reveals that it is the winter winds that are bringing the forest to its knees. It is not the cold winds that kill the branches but the warm, drying winds of late winter. These winds melt the branches' thin protective coats of frost and snow and then draw water out of exposed needles. The tree cannot replace this water, since its roots are frozen into the rocky soil. Below timberline, the trees wear thick coats of snow through the winter that protect them from the drying winds.

Freezing temperatures can prune back trees as well, but not usually in winter. Timberlines in the southern Interior experience balmy winters compared with those of northern forests. The real harshness in southern timberlines occurs in summer, which is cool and very short. The mean July temperature at an Interior treeline is less than 10° C, compared with over 16° C in a far northern valley forest. In addition, there are only about one hundred frost-free days in a year at this altitude, and usually the longest period between two frosts is a mere three weeks.

Trees at timberline have little time, therefore, to put out new growth, and seeds have little time to germinate. The critical period for new tree growth is in the last half of summer, when an early frost could easily kill the soft, immature needles. Mature needles can withstand all but the lowest winter temperatures imaginable. They have a thick wax layer to prevent water loss and a low water content to minimize the effects of freezing. But even the small amount of water in needles can freeze and do damage to cell structure if bitterly cold weather hits before the tree is prepared.

Many trees in northern forests can withstand temperatures as low as -40° C through a process called supercooling. Water won't freeze at 0° C if it has no minute particles or rough edges to initiate crystal formation; if cell walls are smooth, the water in them can supercool to -40° C before ice begins to form. But the hardiest trees—those found in the far north and near treeline on mountains—use a different strategy to survive even more frigid temperatures. It is the formation of ice crystals within cells that quickly kills vegetation by causing cells to rupture and die. Extremely hardy conifer species can move water out of cells and into intracellular spaces where ice can do little damage, a process triggered by increasingly severe frosts. Thus, it is a sudden, hard frost in early August that can affect new twigs. Their water-filled needles, still soft to touch, are quickly killed back.

Pages 12–13: **The mountain behind Lower Tuchodi Lake in Northern Rocky Mountains Provincial Park shows the clear treeline separating forest and alpine tundra.**

Simply put, treeline in the southern Interior mountains is at the altitude where the summer is too cool and short for trees to harden new growth and where winter winds are so strong that they dry out exposed twigs. Stunted mats of krummholz can survive at higher elevations because some of their new growth is insulated from the winter winds by the snowpack. The krummholz line, above which no conifers grow at all, marks the altitude where the growing season is too short to allow germination of trees.

Northern Timberlines

Across northern British Columbia, the timberline zone is somewhat different from that of the southern mountains. Instead of flower-filled meadows between clumps of conifers, the area above the forest line is a mosaic of conifers, mostly White Spruce, and a dense growth of willow and birch shrubs. The willows and birches continue up-slope above the limit of the conifers, eventually giving way to the treeless alpine zone.

Conifers usually have an advantage over deciduous plants in areas that have short, cool summers. Although their needles are relatively inefficient, they can photosynthesize on any winter day that is warm enough and

are fully leaved when spring arrives. At high altitudes in the far north, however, the long summer days apparently favour the production of deciduous leaves, which are more efficient photosynthesizers and are shed before the deadly autumn frosts.

Thus, the forest line of northern British Columbia may be governed simply by low temperatures. The spruce grows as far up the mountains as the autumn frosts allow, and the willows and birches continue up to the point where the growing season is too short to allow germination. The transition between the dense willow-birch shrub zone and open alpine meadows (perhaps *shrubline* would be an appropriate term) is often very abrupt, suggesting that this boundary replaces the krummholz line of southern mountains.

Below the Treeline: The Subalpine Parkland Transition
The zone between the forest line and treeline is one of the most beautiful forest types in British Columbia, a favourite with hikers for its stunning combinations of dark, fragrant trees, spectacular flower blooms and open vistas of the mountains beyond. These open parklands, like the treeline itself, are the result of several forces. The importance of each force varies with geography and climate.

Tree germination is so difficult and early growth of seedlings so slow at these altitudes that any factor that can clear the forest or further suppress germination can produce meadows in the midst of forests. You can measure the slow growth of small trees yourself next time you are in one of these meadows. Find a small conifer a metre or so high and count the distinct whorls of branches up the trunk. Every year, conifers produce an erect shoot, the "leader," from their terminal bud, as well as a circle of side branches. Small trees in open areas retain most of these side branches for years, so it is relatively simple to count the whorls and come close to the tree's true age in years. Trees near sea level in British Columbia can easily grow a metre in a couple of years, but at the forest line it can take a fir fifty years to reach its first metre and another century or two to produce a forest.

Pages 16–17: **Many subalpine meadows in the southern Interior of British Columbia are carved out of the high forest by fire, as the weathered logs on this ridge in the Monashee Mountains suggest.**

The Creation of Subalpine Meadows

In the barren rocks of Castle Peak in the southern Chilcotin, 130 metres above the present-day treeline, a group of weathered stumps bears witness to a time when forests grew higher on these mountains. Local climates are always changing—over millennia, centuries or even mere decades. The warmest period since the glaciers retreated from British Columbia occurred between nine thousand and six thousand years ago. Since then, the climate has been more or less the same as it is today, except for a cold period between 1600 and 1900—the Little Ice Age. Cold winters and heavy snowfalls during the Little Ice Age likely destroyed many timberline forests, their trees already growing at the limit of endurance.

In the last century, trees have probably been inching their way up the mountains again. This upward progress, like the downward destruction of the Little Ice Age, does not usually proceed in nice horizontal lines. Instead, trees invade meadows in patches—long stringers moving along relatively fertile soils or up snow-free ridges; tree islands forming around hardy pioneering individuals. Thus, some of the subalpine meadows we

see today could be remnant alpine tundras now surrounded by forests about a hundred years old.

The present-day climate creates meadows in subalpine forests as well. Just as a heavy accumulation of snow lowers the treeline on coastal mountains, heavy snow accumulation in bowls, in gullies and on the lee side of ridges can make it almost impossible for trees to become established. These snow-created meadows are usually carpeted with mountain-heathers and sedges that tolerate very short growing seasons. Cold air pooling into mountain cirques and northern valleys can also shorten the growing season to the point of excluding trees.

The Role of Fire

One hot July weekend we were hiking along a subalpine trail in the western end of Manning Park. The flowers were in full bloom—meadows knee-deep in valerian, lupine and paintbrush. Towering clouds built up in the afternoon, and as we reached the last ridge we were suddenly enveloped in a violent thunderstorm. We huddled in a small valley as a series of lightning flashes lit up the dark fog and deafening thunder echoed off the rock walls. An hour later it was over, mist rising off the green meadows, puddles of hail melting in the sun and, here and there, plumes of smoke rising from distant ridges where lightning strikes had ignited trees. It had been an exhilarating and frightening experience of one of the forces that had created the flower meadows we had enjoyed all morning.

Many Interior subalpine meadows are born of fire, the flames of lightning strikes clearing thick forests of spruce and fir. Each meadow is only temporary (on the time scale of a tree's lifetime), slowly regrowing to forests of Subalpine Fir and Engelmann Spruce, but the patchwork of meadows is permanent. In the drier parts of the southern Interior, fires return to any given subalpine forest about once every hundred years, so by the time one meadow begins to disappear beneath the branches of young trees, another is cleared.

Coastal Subalpine Meadows

Subalpine meadows on coastal mountains are almost all snow meadows dominated by heaths and sedges; colourful fields of flowers are generally small and restricted to seepage areas and stream banks. Masses of tiny pink, cream and white heather bells brighten the meadows in summer, and in fall the huckleberry and blueberry leaves turn yellow, orange and scarlet against the dark green heather needles, providing a beautiful backdrop for berry pickers.

Mountain Hemlock

Amabilis Fir

Yellow Cedar

As you climb higher in the Coast Mountains, the meadows become larger and the forest is isolated into small islands of Mountain Hemlock, Subalpine Fir and Yellow Cedar. Amabilis Fir, often common in the forests just below forest line, is rare in this parkland zone. Some meadows show signs of invasion by seedling hemlocks. These small trees germinated during a series of warm, dry years when low snowpacks melted early. They grow exceedingly slowly, but after they reach a critical height their crowns will be above the snow early enough to actually speed snow melt around them. Because the branches are dark, they readily absorb sunlight and release the energy as heat. This phenomenon significantly increases the growing season for these small trees. It can also save them from serious damage from the snow mould, *Herpotrichia juniperi*, that forms black mats on branches that remain under the snow for any length of time in spring.

The change from a moderate coastal climate to a more continental Interior climate occurs over a short distance as one travels through the Coast Mountains, and mountain meadows on the eastern slopes are similar to those in Interior mountains. Even as close to the coast as Garibaldi Lake, the heather meadows are broken by wide swaths of lupine, cinquefoil and valerian, and Subalpine Fir is an important member of the high forest.

Below: A female Blue Grouse on a patch of late-melting snow.

Opposite page: The Vancouver Island Marmot is much darker than its mainland relative, the Hoary Marmot—the former has a dark, chocolate-brown coat with a white nose and a white streak along the belly.

Blue Grouse are common in these meadows, the females leading downy chicks that tumble over the heather. The female is sometimes also followed by a male with amorous intentions, its black tail raised and fanned and neck sacs inflated, exposing circular patches of bright yellow skin surrounded by startlingly white feathers—looking for all the world like a pair of fried eggs, one on each side of the neck. This display is accompanied by a series of deep hoots amplified by the inflation of the air sacs. Males usually hoot up in trees, where they can be almost impossible to find; they are ventriloquists, and Mountain Hemlock foliage is very dense. If a female comes by, though, the male takes to the ground to pursue her. After mating, males take no part in nesting or the care of young but spend all spring and summer hooting and hoping.

One of the rarest mammals on the continent, the

Vancouver Island Marmot is found only on a few mountain ridges on central Vancouver Island. Fewer than five hundred are known to be alive today, and most of the known population is restricted to a single, long ridge complex between Green Mountain and Gemini Mountain, southwest of Nanaimo. There they live in lush subalpine meadows on steep mountain slopes, where avalanches and snow creep limit tree growth and where food plants such as lupine, Indian-paintbrush, meadow-rue, Cow-parsnip and Woolly Eriophyllum thrive.

No one knows how long these marmots have been isolated from their mainland cousins—they may have colonized the island since the glaciers left ten thousand years ago, or they might have survived the Pleistocene ice ages in an ice-free refuge, perhaps along the island's west coast. Marmot bones seven hundred to over nine thousand years old have been found in caves to the north and west of the present range, suggesting that the species may have been more widespread in the past. In fact, the species was first discovered by zoologists on Douglas Peak near Port Alberni, a mountain where no marmots have been seen for over thirty years.

Why is this species so rare? Part of the answer may be that lush mountain meadows are not common on Vancouver Island and are separated by great swaths of dense, dark forests. The colonies that survive in small patches of ridge-top meadows may die out occasionally, and recolonization is chancy across the wide, deep valleys. Young marmots do disperse through these forests, but most do not find good alpine habitat on the other side. They are also easily hunted, and their limited range may have made local populations susceptible to elimination by early hunters.

One habitat that marmots have taken to with gusto is clearcut logging slash. Because slash is an extremely common habitat on the island, this predilection would seem a great boon to marmots. But some scientists are worried about this trend. First, clearcuts are destined to become thick, regenerating forest in a few years, the antithesis of good marmot habitat. Second, preliminary research has shown that the colonies in clearcuts seem to be unstable; that is, the turnover of individuals in them is very high. Does this high turnover mean that a great number of them die attempting to hibernate in the clearcuts, or do they keep dispersing to new, greener pastures? Do clearcuts act as sinks, siphoning off the young marmots dispersing from nearby colonies, while colonies on distant mountains die out? Only further studies will answer these questions and help ensure that this unique British Columbia species prospers in the wild mountains of Vancouver Island.

Southern Interior Subalpine Meadows

These subalpine meadows are most famous for their fabulous fields of flowers, blooming synchronously in the short mountaintop summer. These flowers appear in two distinct "bloom waves." The first bloom wave occurs immediately after snow melt. This wave is usually at its peak in June, depending on the altitude and the year, but its representative flowers can be found later in the summer alongside any late-melting snowbank. These early flowers are species with bulbs or tuberous roots, which allow rapid growth and flower production before leaf growth. The flowers are usually borne on short stalks to avoid the snow-cooled winds and produce a carpet of ankle-high white, cream and yellow blooms.

The starchy bulbs of the Yellow Glacier Lily were an important part of the diet of southern Interior Native people. They were gathered in late summer and cooked by steaming, baking or boiling.

The showiest species is the Yellow Glacier Lily, which sometimes covers entire mountain ridges in a mantle of gold. Another conspicuous early flower is the Western Anemone, whose large cream flowers grow close to the ground on short, hairy stalks. Anemones are also known as windflowers; *anemos* is the Greek word for wind. After the anemone flowers are pollinated, the stalk grows 20 to 30 centimetres and produces a distinctive moplike seed head, often called a tow-head baby. The long stalk puts the seed head above the taller flowers of late summer so that the wind can spread the long-tasselled seeds.

The third important member of the first bloom wave is the tiny Spring Beauty. Like the lily and anemone, it can also occur in single-species concentrations, covering whole meadows in a pinkish-white mass of flowers. Another name for it is Indian potato, a reference to the small white corms on its roots, which are harvested by the Interior Native people. They are also a favourite of Grizzly Bears, which often dig up large areas of plants with their long claws.

Often the first snow-free areas in mountain meadows are along small brooks overflowing with meltwater.

Here White Marsh-marigolds and Globeflowers grow and bloom in the shallow flowing water. Both species have white flowers similar to those of the closely related Western Anemone, but their leaves are quite different.

As the snow melts and the soil surface dries in midsummer, other species of plants grow and mature. By late July, these plants are ready to flower, and the second bloom wave covers the subalpine meadows. This is a much more diverse bloom than the first, both in colour and form, and includes white valerian, red paintbrush, blue lupine, yellow arnica, avens and ragwort, and purple daisies and asters.

Take a close look at the paintbrush flowers—the red or orange "petals" aren't petals at all but brightly coloured leaves. The true flowers are the green shoots coming out at the bases of the leaves. The red or orange colour is a clue that paintbrush is a hummingbird flower—one that is pollinated almost exclusively by hummingbirds rather than bees or flies. The long, tubular flowers are the

Summer in subalpine meadows is a brief but spectacular blaze of colour as Arctic Lupines, Common Red Paintbrushes and other mountain flowers hurry to set seed before the late-summer frosts.

right shape to accommodate hummingbird bills; no honeybees and few other insects have tongues long enough to get at the nectar at the base of paintbrush flowers.

Indian Hellebore

In seepage areas the large leaves of Indian Hellebore are distinctive. Although the small green flowers of the hellebore aren't striking to look at, their delicate watermelon fragrance is worth sniffing. Don't be tempted to add the leaves to a wilderness salad; the entire plant is extremely poisonous.

Although many of the trees found at timberline in the mountains of the southern Interior are also found below in the subalpine forests, two conifer species are more or less restricted to timberline itself—the Whitebark Pine and the Alpine Larch.

Whitebark Pine is easily identified by its upsweeping branches, often multi-trunked form, bright green needles clustered five to a bunch (compared with two in the Lodgepole Pine, a related species occasionally growing near timberline), smooth grey bark and spineless, purplish cones. This species is common on relatively warm, often south-facing slopes in open parkland and can be found as close to the Pacific as Garibaldi Lake in the Coast Mountains.

White Marsh-marigold

Western Anemone

Globeflower

Buttercups versus Buttercup Eaters

The leaves of the Western Anemone and Globeflower, like the leaves of many of their relatives in the buttercup family, contain ranunculin. This is an innocuous chemical itself, but it readily hydrolyzes to form an extremely irritating oil, protoanemonin. This chemical blisters the tongue and throat if eaten, an obvious deterrent to grazing animals. Plants in the buttercup family have long been used in folk medicine to raise hot blisters when applied as poultices. Cultivated mustard, which is also used in poultices, produces similar irritant oils.

The Whitebark Pine has a remarkable relationship with the Clark's Nutcracker. The Clark's Nutcracker is member of the jay and crow family and, like its cousins, is an opportunistic feeder with a predilection for robbing other birds' nests to feed its own young. And like the Gray Jay, it can become quite a panhandler for food, especially in mountain parks such as Manning and Cathedral Provincial Parks, and throughout much of the Rocky Mountains. But in late summer and early autumn, the Clark's Nutcracker switches to eating large pine seeds, which it finds in the cones of Ponderosa and Whitebark Pines.

Whitebark Pine

Alpine Larch

The Whitebark Pine is not an ordinary pine. Most pines have thick cones armed with nasty spines or a heavy coating of pitch to deter squirrels and other animals from eating the seeds, which are usually small with large wings for flying far on the fall winds. Whitebark cones, in contrast, have no spines or heavy pitch and look like dark purple flowers on the ends of branches. The seeds are very large and wingless, and the cones do not open fully to release the seeds when mature—if the cone did release the seeds, they would just fall to the ground and grow up competing with the parent tree. Instead, Whitebark Pines rely almost entirely on Clark's Nutcrackers to open their cones and gather, disperse, and plant their seeds.

The nutcrackers have a throat pouch with enough space to store up to 100 pine seeds. When it is full they fly off, sometimes for several kilometres, to a suitable place to cache the seeds for winter food. A suitable place is usually a windswept, southern exposure with sandy soil, allowing the birds easy access all winter to their supply of seeds. It's no coincidence that this habitat is also ideal for Whitebark Pines. The birds land on the ground and bury the seeds in groups of five or six under a couple of centimetres of soft soil; then they take off and harvest more. An adult nutcracker will cache 30,000 to 100,000 seeds in this manner each fall!

How do they find them again? Studies have shown that Clark's Nutcrackers have remarkable spatial memories and outperform zoology graduate students time and time again in tests of finding buried seeds. They don't find all 30,000 seeds, of course, and come summer the hillsides are dotted with clusters of tiny pine seedlings, a half dozen or so in each group, depending on how many seeds the nutcracker planted. Some of these clusters survive

Rufous Hummingbirds use the flower-laden mountain ridges as their migration highways en route to their winter quarters in the Sierra Madre of western Mexico.

Hummingbird Migration

By the time the subalpine flower meadows are in full bloom in late July, male hummingbirds are moving south. They need to refuel regularly, and one of their favourite fuels is paintbrush nectar.

An old wives' tale has it that hummingbirds migrate south on the backs of geese, but that story has two fatal flaws. One is that hummingbirds migrate before the geese—even females and young have left by early September, whereas geese don't pass through southern British Columbia until early October. Another is that geese migrate to the central and southern United States, settling in for winter well north of the Mexican winter home of the hummingbirds.

Clark's Nutcracker

to grow into the many-trunked growth form so typical of Whitebark Pine.

Alpine Larches are often the last trees one encounters on a hike up a mountainside in southern British Columbia. Their deciduous habit likely gives them the advantage over evergreens in surviving bitter frosts in late fall, but they need full sun in summer to get the energy to make all those new needles. Alpine Larches can only germinate in cool conditions, though, and are therefore commonest on open, north-facing slopes, nicely complementing the Whitebark Pines growing on south-facing slopes. Alpine Larches grow in such harsh conditions that successful seed production and germination is rare; it is an uncommon event to find a young seedling.

Whitebark Pines and Alpine Larches are often the pioneer members of tree islands. These clumps of trees form in alpine meadows after one individual grows tall enough to shelter the growth of new seedlings. These islands can gradually grow outward as more trees germinate around their circumference. The pioneer tree may eventually die, leaving a hollow centre in the island. As more central trees die and new ones grow outward, the island can become a doughnut-shaped "tree atoll."

If the pioneer tree of a tree island was a Subalpine Fir, the younger trees are likely genetically identical copies of it—clones—since this species often reproduces by layering. In this process the lowest branches, when pressed to the ground by the spring snowpack, put down roots and send up trunks of their own, which eventually become separate trees.

Columbian Groundsquirrels are true hibernators, spending the winter in their underground burrows in a state of deep torpor, their body temperatures the same as that of the air around them.

Many of the small mammals of subalpine meadows spend much of their lives underground, for the most part to escape predators and the rigours of the mountain climate. The groundsquirrels and marmots, for example, sleep underground most of the year, coming

above to feed only on summer days. One animal that you rarely see above ground in daylight is the Northern Pocket Gopher, a curious rodent that tunnels like a mole. Unlike moles, however, it eats plants instead of worms and insects. If you see "molehills" in the southern British Columbia Interior, they are the work of pocket gophers; molehills on south coast lawns are the real thing. Although they have a bad reputation in Interior vegetable gardens and alfalfa fields, pocket gophers provide valuable roto-tilling service to subalpine meadows, aerating the soil and increasing water absorption so that the surface run-off doesn't all disappear downhill. North of the Thompson-Shuswap valley, this service is provided to some extent by other rodents, including lemmings, voles and groundsquirrels.

Northern Pocket Gopher

Below: Pocket gophers remain active all winter, packing excavated dirt into snow tunnels on the surface of the ground. These dirt-filled tunnels are evident in summer as raised, miniature highways of earth running through the meadows.

Pages 32–33: The Alpine Bearberry, a northern relative of Kinnikinnick, turns the tundra crimson as its leaves turn colour in late August.

Northern Timberline: The Shrub Tundra

The subalpine zone in northern British Columbia (north of Williston Lake) is characterized by tall willow and birch shrubs. Although these shrubs can be difficult to walk through, hikers resting in small openings in the cover are rewarded with flowers typical of the spruce forests below, such as Tall Bluebells, ragwort, larkspur, monkshood and shootingstars. As you break out into the open alpine zone, the ground is carpeted with blueberries and Alpine Bearberries.

Northern subalpine valleys often have two timberlines—one on the mountain ridges and another just above the valley bottom. These lower timberlines are found in wide valleys where heavy, frigid air collects in the bottomlands. This ponding of cold air significantly lowers temperatures throughout the year and can often result in permafrost conditions in the valleys. Such valleys are filled with shrubs, grasslands and fens and provide the only breeding habitat in the province for Gray-cheeked Thrushes and Smith's Longspurs.

Snowshoe Hares don't often venture beyond the cover of the spruce forest, but Arctic Groundsquirrels are common in the alpine meadows of the far northwest and the runways of Tundra Voles and Brown Lemmings can easily be found in grassy swales. Almost everyone is familiar with the story of lemmings—how entire populations go on crazy migrations, eventually committing suicide by jumping off cliffs into the sea. That's the Walt Disney version, though; the real story is somewhat less dramatic but no less fascinating.

Lemmings are large, tundra-dwelling mice with short legs, short ears and almost no tail. There are several species around the Northern Hemisphere; the one in British Columbia is the Brown Lemming, distinguished by its orange backside. It is found in the high mountain tundras of the northwestern quarter, coming as far south as Tweedsmuir Provincial Park.

Most lemming populations, as well as those of their close relatives, the voles, or meadow mice, undergo tremendous peaks and crashes, usually at three- to five-year intervals. These population cycles have puzzled ecologists for over fifty years, and the cause of these cycles is still somewhat of a mystery. The following is a simplified account of what seems to happen.

Brown Lemming

During the low phase, lemmings are very hard to find. Over the next year or two, lemming predators—owls, hawks, foxes and weasels—are forced to disperse from the tundra and switch to other prey species if they can. The stage is now set for the increase phase. Given a deep, insulating blanket of snow, lemmings can breed throughout the winter as well as in the summer, increasing their numbers tremendously in one year. Soon their populations are in plague proportions, and their aggressive instincts take over.

Lemmings are highly territorial, and at high population densities young animals are constantly being driven out of adult territories. At this point lemmings appear to be everywhere, and young animals are seen dispersing through unsuitable habitats such as forests. And yes, they can even be seen attempting to swim across large lakes, desperately trying to find a place where their rear ends aren't being constantly bitten by bigger lemmings.

Back on the tundra, predator numbers are up, hawks and owls taking advantage of the lemming feast by producing large, healthy families. The adult lemmings,

Afternoon sun along the Haines road highlights thickets of willow and birch shrubs, which dominate the subalpine zone of northern British Columbia.

perhaps distracted by all that aggression, produce fewer and fewer young and begin to die of old age themselves. Within a year, the lemming population has all but disappeared.

THE ALPINE TUNDRA

As you approach treeline, the krummholz skirts of trees grow larger and the erect trunks shorter. The last trunks are mere krummholz flags: short poles with a tuft of branchlets near the top on the leeward side of the trunk. The lower branches above the snowpack are scoured off each winter by blowing ice. Finally, you reach an altitude where there are no trunks at all, only krummholz mats huddled behind rocks. In this cold, wind-lashed environment, plants cannot grow large enough to create more favourable microclimates, as the forest does a few hundred metres downhill. Rocks and

natural hollows are often the only sites that offer enough protection from the wind to allow plants to germinate. Krummholz trees growing behind rocks often spread gradually leeward, new growth surviving only in the wind shadow of the tree itself.

Alpine Plant Life

Tundra plants reward close inspection; you almost have to get down on your knees to appreciate them at all. Dwarf willows lie flat against the gravel except for their upright, down-jacketed catkins. A hurrying hiker could mistake Spotted Saxifrage for a moss with tiny white flowers; a face-to-face look reveals the exquisite scarlet and gold spots on the petals. Alpine Forget-me-nots reflect the blue mountain sky with their tiny flowers. Like the forget-me-nots, many of the tall herbs of the subalpine meadows are replaced by prostrate relatives above treeline—the catchflies by Moss Campion, Arctic Lupine by Dwarf Mountain Lupine, Subalpine Daisies by the tiny Golden Fleabane.

Gravelly ridges on mountaintops are often covered in White Mountain-avens; the cream flowers appear shortly after snow melt, replaced in late summer by a carpet of tasselled seed heads.

It may be hard to think of alpine meadows as deserts, but after the snowpack is gone, water is very hard to come by on high, rocky ridges. Summer rainwater disappears quickly into the shallow, well-drained soil, and the intense sunshine and dry winds quickly desiccate leaves. Alpine plants therefore share many characteristics with desert plants. Some, such as stonecrop, are succulent and fleshy, enabling them to store water through the summer drought. Others, such as Sulfur Buckwheat, locoweed, pussytoes and Dwarf Mountain Lupine, have leaves encased in grey, woolly hairs to reduce evaporation.

Many alpine rocks are colourfully patterned with living patches of bright orange or vivid yellow-green

The Cost of Sex

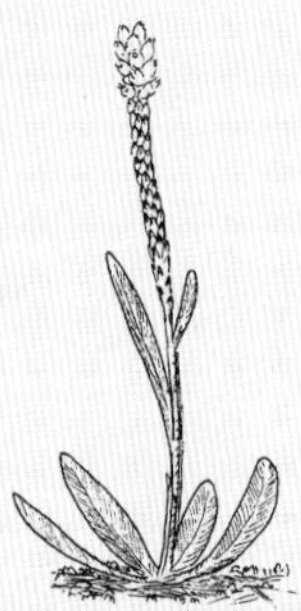

Alpine Bistort

Alpine plants live in an uncompromising environment where any extra costs could mean the difference between surviving and perishing, and one of the most metabolically expensive practices in plants and animals is sex. The cost of producing a new plant through sexual reproduction—which requires flowers, pollination, seed production and seed survival—is about ten thousand times that of producing a copy through a runner, as strawberries or buttercups often do. That's why lawns are often overrun with highly successful buttercups. It's not surprising that many alpine plants choose this method of copying themselves.

Sex does have great advantages, though, so most species practise both methods of reproduction. The Alpine Bistort produces a flower stalk with small white flowers at the top and what looks like green flower buds below. The buds are actually vegetative bulbils, which can also grow into new plants after they drop off the stem. Other species make sex more efficient by being dioecious—producing the male and female flowers on separate plants. Some of the most successful alpine plants fall into this category—the willows, Moss Campion, Mountain Sorrel and Cloudberry.

and black—crusts of life covering the stones. These are lichens, and they can tell much about the area they grow in. The circles of Rock Orange Lichen, for instance, grow only where there is a source of calcium or nitrates, often from bird droppings or mammal urine. These brightly coloured rocks therefore indicate favourite lookout sites for pikas, cliff nest sites of hawks and eagles or song posts of Horned Larks. The dark brown mats of Blistered Rock Tripe and black tufts of Velcro Lichen grow on acid rocks.

At first glance, the alpine tundra flower communities of British Columbia seem to violate one of the tenets of ecology, that species diversity decreases as you get closer to the poles. Yet alpine flower communities on mountains in southern British Columbia have fewer species than those in the north. The same pattern holds true for alpine butterflies. This odd situation is easy to understand if you picture the alpine habitats of the south as increasingly small, isolated islands of tundra in a sea of forest.

The alpine meadows of northern British Columbia are much larger islands, all close to other alpine meadows and closer to the huge arctic tundra that was the source of the alpine species. The farther away from the source populations the southern tundra islands are, the fewer arctic emigrants there are that have been able to reach them.

When species disappear from small, isolated islands for one reason or another, they are rarely replaced by new ones, since the source populations are so far away. And the smaller the island, the smaller its plant and animal populations, and it is all the more likely that these populations will disappear. In the north, species don't often disappear from the large tundra islands, and seeds can easily blow or butterflies fly from one mountain ridge to another, providing a constant supply of species to replace any that disappear.

Insect Singles Bars and More

Sarcophagid fly

Tachnid fly

Have you ever hiked up a mountain on a fine summer morning and plunked down at the top, relaxing on the warm rocks, beaming with accomplishment—and found yourself surrounded by buzzing flies? I don't mean the biting kind, just all sorts of flies: big ones and small ones, yellow ones and black ones, hovering ones and crawling ones. If you don't remember such an event, look closely the next time you are on a mountain or hilltop. For a moment, take your eyes off the magnificent landscapes below and focus on the rocks around you. If there is a cairn on top, watch that. If it is warm and sunny, you will see the singles bar of the insect world!

Most male insects go about finding mates in conventional ways—they hang around the place they were born, they sniff the air for a particular perfume, they check out the best local flowers, or they search the places where females may want to raise a family. Some insects, however, have come across a simple but ingenious way of getting together. These are generally widely dispersed but strong-flying insects, primarily true flies and some butterflies and moths. Finding each other in the big, wide forest may be a challenge for these insects, so they have evolved an intuitive understanding—"Let's meet at the top of the hill." All they have to do is fly uphill until there is no more hill to fly up.

Generally speaking, males fly to hilltops and stay there, waiting for females. The females fly up for brief visits only—they quickly find a mate and then journey back down to their particular habitat on the hillside or in the valley, where they lay their eggs. Watch quietly from your hilltop resting spot and you will see bristly tachinid flies zooming around in frenetic haste; flower flies, sarcophagid flies and blow flies maintaining their watch from the warm rocks; and swallowtail butterflies flying rapidly over the ridge and down the other side a ways before resuming their upward flight. Hilltops are also just about the only place you will ever see a male horse fly—you can distinguish males from their female counterparts in two ways: by their extra-large eyes that meet across the top of their heads and by the fact that they totally ignore you!

Some insects, especially some ladybird beetles, meet on the tops of mountains to gather into large hibernation groups. In the late summer, they can be seen swarming

Old World Swallowtails, like many butterflies and other insects, meet on hilltops to mate.

over mountaintop boulders before they crawl beneath them to spend the winter in a mass of friendly red bodies. In the British Columbia Rockies, bear biologists have discovered that some Grizzlies abandon good berry crops on the slopes to go to the peaks and fatten up on beetles and moths.

Dead insects can be surprisingly numerous on the vast snow fields of the higher mountains. These are unlucky members of the aerial plankton, insects that disperse by flying high into the atmosphere and letting the wind take them away from their birthplaces. The wind currents rise up over the mountains and then are cooled by the altitude, ice and snow; the resulting downdrafts deposit the insects on barren rock and chilling snow. Most are flying ants, whose presence over the mountaintops can be detected by flocks of chattering Black Swifts. These large, swallow-like birds specialize in consuming swarms of ants, flying long distances to the mountain peaks each day from their nest sites, usually located in moist canyons far below.

Black Swift

Birds and Mammals

I (Richard) once led an autumn field trip for university students into the Cascade mountains to study the animals of the treeline. One of our objectives was to find White-tailed Ptarmigan, the small grouse of alpine habitats in western North America. When we paused for lunch, a student handed me a small white feather. "Ptarmigan?" he asked. I nodded encouragingly. "Is that ptarmigan poop?" he asked again, pointing at a pile of toothpaste-shaped droppings. I nodded again. "Are those ptarmigan?" he continued, indicating a dozen lichen-coloured grouse just now getting to their feet and trying to walk out of the lunch circle without being noticed. I had long known how easily you could walk by a well-camouflaged ptarmigan, but to half finish my lunch within a few feet from a dozen was too much!

There are actually three species of ptarmigan in the world, all of them found in British Columbia. They are distinguished from other grouse by their heavily feathered feet and white winter plumage, both adaptations to their life in the snow. The largest is the Willow Ptarmigan, found throughout the willow shrub tundra of northern British Columbia. The Rock Ptarmigan is a moderate-sized species found in grassy alpine habitats in the north and throughout the Coast Mountains. The White-tailed is the smallest and most widespread species, found in rocky alpine areas throughout the province.

In fall, ptarmigan moult into their white winter plumage, Willow and Rock retaining their black tails (usually hidden under long white covert feathers) and White-tailed turning totally white except for black eyes and beak. The white plumage not only makes them hard to see against the snow but also keeps the birds warmer. White feathers (and white hairs for that matter) are hollow and empty of pigment and are significantly better insulators than coloured feathers.

In late fall, young ptarmigan often leave their mountaintop homes to find new ridges to colonize and occasionally turn up in very odd places. A friend of ours was surprised one November afternoon to see a White-tailed Ptarmigan strolling across the postage-stamp-sized lawn of her townhouse on the Fraser Delta.

Come spring, the females moult back into their cryptic summer plumage, but the males delay the moult for a month or so to perform their courtship rituals in

white winter dress. Willow Ptarmigan males actually regain their reddish brown head and neck and then stop the moult to jump and cackle in a startling two-tone spring outfit.

The barren ridges above the treeline attract a small but distinctive songbird community. Alpine tundras in the far northwestern corner of the province are the summer home of the Snow Bunting, a handsome white sparrow with black trim, known to other areas of British Columbia only as a winter visitor. Elsewhere in the province, there are really only three species of songbirds that nest above treeline. Steep slopes, especially those with lush growth around seepages, are habitat for American Pipits, and the ridges themselves are inhabited by Horned Larks. Both the larks and the pipits sing their territorial songs while in flight, a habit common among birds of treeless environments. The third songbird species here is the Gray-crowned Rosy Finch, a chocolate-brown bird with bright pink wings and rump found in the rocky cliffs and ice fields of the highest mountains. Rosy finches

Opposite page: **Probably the best time of year to look for White-tailed Ptarmigan is in August and September, when the lichen-coloured birds are in family groups, making it less likely they can slip by unnoticed. Listen for the clucking calls of the females to their chicks.**

Below: **Horned Larks almost always tuck their nests into the lee of a rock or small alpine plant. The bright orange mouths of the nestlings, conspicuously spotted with black, provide a clear target for the adults at feeding time.**

can often be seen in summer foraging on the permanent snow fields, picking up windblown insects and seeds.

Almost all the birds of the alpine tundra abandon the mountaintops in winter for milder climates. Larks fly off to southern grasslands, pipits to temperate coastal mud flats and Gray-crowned Rosy Finches to gravel banks of intermontane valleys. Only the ptarmigan stay, huddled under the snow for up to 80 per cent of the day, feeding on seeds on the windswept ridges.

The less mobile mammals, however, are forced to stay and deal with the long, frigid nights and short,

Migrating Hawks and Songbirds

The next time you are hiking along a windswept mountain ridge in August or September, look up. Chances are a hawk or an eagle will be somewhere in view, sailing along on the updrafts created by the prevailing winds deflected upward by the ridge. These birds are on their way south, like a lot of birds at that time of year, and are using the ridges as expressways. In parts of the world where mountain ridges are not as common as in British Columbia, remarkable concentrations of hawks and eagles are seen in spring and fall. The largest concentrations of this sort in British Columbia are along the headlands west of Victoria, where the southbound birds run out of mountain ridges and are forced to wait for good soaring weather before crossing the Strait of Juan de Fuca.

Cypress Provincial Park is a good subalpine site near Vancouver, as is the Three Brothers area of Manning Park, but almost any good ridge will provide some action. Try looking on days when there is a steady wind out of the north or northwest.

Mountain ridges are also used by songbirds on their southward migration. They may be attracted by the prospect of a good tail wind, but they are certainly taking advantage of the end-of-summer feast of berries and insects available. Songbirds avoid high mountains during spring migration if at all possible—the deep snow hides any source of food and could prove fatal if the weather changes quickly.

hungry days. To survive, mammals use a variety of strategies, most of which involve lying low and using as little energy as possible. The extreme form of this strategy is hibernation, exemplified by the ground-squirrels and marmots. These mammals gorge themselves all summer on grasses, flowers and other plants, putting on a thick layer of fat. At summer's end they go into their burrows to a special hibernation den, curl up and fall into a deep sleep. This deep sleep is called torpor, and in it the animal's body temperature falls to near that of the surrounding air. In this way the hibernator can save a large amount of energy, since it doesn't have to burn very much fat to keep up its temperature.

Above: **Pikas spend the summer gathering leaves, which they cure in hay piles as their only source of winter food.**

Pages 48–49: **The white tufts of cottongrass beside the Ptarmigan Lakes in Tweedsmuir Provincial Park are a characteristic feature of moist northern tundras.**

Instead of storing winter food as fat, the pikas exemplify the saying "Make hay while the sun shines." These little relatives of the rabbits and hares live on rocky talus slopes, signalling their presence to hikers with a sharp, nasal *enk*! call. Pikas spend the summer gathering grass, herbs and flowers, carrying them back to the rocks near their home burrows and spreading them out to dry. When their hay is cured, pikas move it under the cover of large, flat rocks for use during the long winter.

Some of the smaller rodents can get through the winter by foraging under the cover of snow, eating dried grasses and herbs. The Heather Vole leaves behind evidence of its subnivean lifestyle when the snow melts in summer, exposing a cosy little grass nest close to a large pile of tiny droppings—the mouse's winter outhouse.

PART TWO

THE SPRUCE KINGDOM

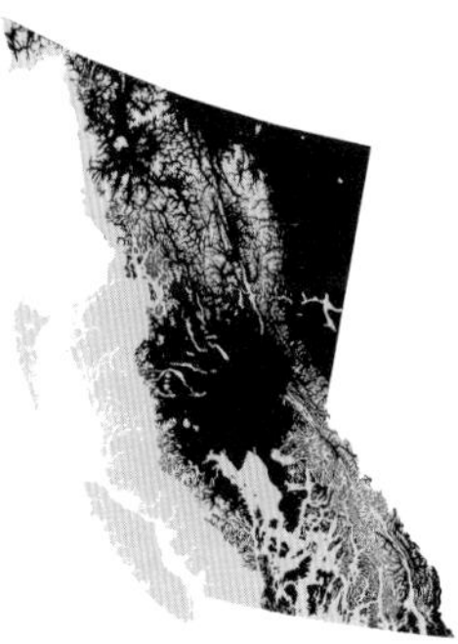

A GREAT BAND of cold, coniferous forests rings the Northern Hemisphere, spreading across central Canada, southern Alaska, the Siberian taiga and Scandinavia. Where mountain ranges cut across the continents, these forests extend south at higher elevations—dark green fingers between alpine tundra and warmer montane forests. Ecologists describe the northern forests as boreal, from the Latin *boreas,* the north wind; the more southerly high mountain forests are generally termed subalpine. Both types of forest are more or less defined by the presence of spruces, with firs interspersed in older stands and large areas of pine in regions recently swept by fire.

The scale of these forests is immense, and to ground-based humans they often appear monotonous and almost featureless. But patterns are everywhere in the spruce forests: spatial patterns caused by fire, frost and water, and the temporal patterns of a continental climate and of dramatic, decade-long population cycles in the animals that inhabit them.

The boreal and subalpine forests look similar to the untrained eye, but there are significant differences in climate and dominant tree species. Although both regions have long winters, the northern plateaus have a more continental climate, a climate of extremes. These plateaus are enveloped in frigid, arctic air for weeks at a time each winter but enjoy long, hot summer days. Winters in the southern mountains are milder, influenced more strongly by Pacific air masses, whereas summer days there are shorter and cooler than on the northern plateaus.

Pages 52–53: **The spire-like shape of conifers in the high mountain forests is an adaptation for shedding the heavy winter snowfalls.**

FORESTS OF SPRUCE AND FIR

White and Engelmann Spruce

The boreal and subalpine forests of British Columbia are dominated by two species of spruce—those prickly-needled, scaly-barked conifers that thrive in cold climates. White Spruce and Engelmann Spruce are sibling species—closely related species that have evolved from a common ancestor only recently. These spruce apparently diverged when they were separated by the glaciers of the Pleistocene ice ages; White Spruce now dominates the eastern and northern boreal forests, and Engelmann Spruce is common in the high forests of the Western Cordillera. The two species mingle and hybridize on the mid-mountain slopes of southern British Columbia and

across its broad, central plateaus, where it can be difficult to find a pure tree of either type. Even experts label them simply as hybrid spruce. In those areas, spruces above 1500 metres in elevation can generally be safely identified as Engelmann, whereas those below 900 metres are usually White Spruce. The keen naturalist can try to verify this rule by examining spruce cones: Engelmann's are papery and flexible, with wavy-edged and more pointed scales; those of White Spruce are stiffer, with rounded, smooth-edged scales.

White Spruce
Engelmann Spruce

Subalpine Fir

THE BOREAL FOREST IS A FORMIDABLE ENVIRONMENT THAT CHALLENGES ALL THOSE WHO VENTURE INTO IT, AND THAT IS ITS TRUE FASCINATION.
—Cameron Young, *Forests of British Columbia*

Many spruce trees have compact tangles of small branches called yellow witches' broom, the result of an infection by Spruce Broom Rust. The stunted needles on the branchlets are bright gold instead of grey-green and are shed in fall. When the rust matures, a strong, musky smell wafts through the forest and flecks of orange spores cover the needles. Most rusts (like most gall aphids—see sidebar) alternate between unrelated hosts such as Subalpine Fir and Fireweed, Lodgepole Pine and aster, spruce and Labrador Tea. The spores of Spruce Broom Rust blow through the forest, infecting its other host, Kinnikinnick.

Subalpine Fir

The other conifer sharing the spotlight with spruce in the cold forests of British Columbia is Subalpine Fir.

Subalpine Fir is called balsam by foresters, but the real Balsam Fir is a close relative confined to eastern North America. The smell of Subalpine Fir's foliage is the smell of the British Columbia mountains, and a brief hint of its fragrance on a cool breeze evokes memories of past mountain adventures. When we are driving through the mountains, we roll down the window at the top of the pass to relive those memories. From a car, you can also identify Subalpine Firs visually by their tidy, symmetrical appearance and their flat, tiered whorls of branches.

Conelike Galls

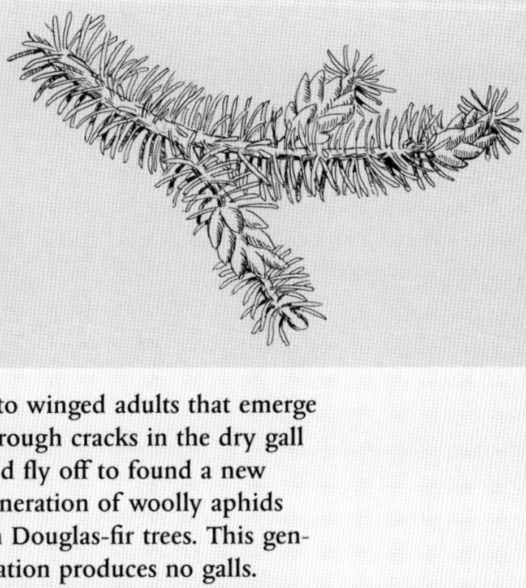

When checking cones, you will probably notice that some at the tips of spruce branches aren't cones at all. These conelike structures are the abandoned homes of the Cooley Gall Aphid, and if you look closely, the scales are really swollen, dried needles. This effect of the aphids superbly illustrates the fact that cones themselves evolved from simpler structures at branch tips.

If you break a gall apart, you will see small chambers at the base of each needle. Each of these chambers held several aphids when the gall was fresh. These aphids started as parthenogenetic eggs—unfertilized but viable, and all female—laid on the soft new twig. The young aphids suck plant juices from the stem, exuding chemicals that stimulate the growth of the gall around them. They later mature into winged adults that emerge through cracks in the dry gall and fly off to found a new generation of woolly aphids on Douglas-fir trees. This generation produces no galls.

The cycle is complete when the progeny of those aphids return to the spruce forests. Cooley Gall Aphids are therefore found only in mountain forests where Douglas-firs are growing nearby. Other species form galls in northern British Columbia, but they either stick to spruce throughout their life cycles or alternate with larch or pine hosts.

The cones of the Subalpine Fir and other species in the "true fir" genus *Abies* are distinctive in a couple of ways. First, they stand erect along the upper branches of the tree, and second, they do not harden and drop off the branches when mature but disintegrate while they are still attached. After the bracts have fallen and the seeds have flown away on the winds, the central cores remain, standing like slender candles atop a Christmas tree.

Opposite page: Although hair lichens grow well in cool, moist forests, they are intolerant of prolonged dampness. The abrupt lower line of these Common Witch's Hair Lichens on the Mountain Hemlock trunks marks the winter snow depth.

Throughout much of the boreal forests, the trees are draped with blackish horsehair lichens. To many people, the lichens make the trees look damaged, sick or even burnt, but they actually do little or no harm. The lichens are merely using the trees as a support to hang in the breeze, absorbing water and nutrients from the air. Like all lichens, they are fungi that have discovered greenhouse agriculture—they farm algae within their own bodies. The fungi provide a moist, nutrient-rich environment for the algae, while the algae use their chlorophyll to turn sunlight, air and water into sugar.

Hair Lichens

Two types of lichens commonly drape themselves on coniferous trees in this region—the cream-coloured witch's hair lichen and the dark brown horsehair lichen. Both are important winter food for Caribou, and both are indicators of old forests. Horsehair lichens break apart easily and spread from tree to tree on the strong mountain winds; snowfields often look like the floor of a barber shop hundreds of metres from the nearest source of lichens. Horsehair lichens cannot tolerate being wet for long, though, and only grow well on open branches typical of older trees; there the wind can dry them quickly after misty rain or snow. Witch's hair lichens are much stronger and don't break so easily, and therefore spread into new forests much more slowly than horsehair lichens.

Its flowers are much smaller than those of its cultivated cousins, but the whorled, coppery-haired leaves of the White-flowered Rhododendron give away its affiliation.

When deep snow blankets the ground, Caribou in the mountains of southern British Columbia depend on these tree lichens for their winter sustenance.

Shrubs and Flowers

A forest, of course, is much more than a collection of trees, and spruce forests are home to many other plants. The understorey of the forest can tell you a lot about the local conditions.

The commonest shrubs of the boreal forest are Highbush Cranberry and Prickly Rose; in moist, rich sites these are joined by jungles of Black Twinberry and Red Swamp Currant. Northern woodland flowers are abundant in mature, open forests—look for carpets of the rounded leaves of Twinflower and get down on your hands and knees to smell the delicate, bell-like flowers. Bunchberries indicate areas of nitrogen-poor soil, whereas the delicate runners of Trailing Raspberry indicate moist, nitrogen-rich sites.

In the mountain forests of southern British Columbia, Black Huckleberry, White-flowered Rhododendron and False Azalea are important parts of the understorey. The rhododendron's branches sweep downhill to lie flat under the heavy burden of winter snow, but to any hiker trying to bushwhack uphill in the summer they can be an impenetrable tangle, giving this shrub

Of Moose and Moss

One of the most remarkable plant-animal interactions in the boreal forest involves a strange group of mosses in the genus *Splachnum.* In an environment where nutrients are hard to find, these plants have come up with a novel solution—they grow almost exclusively on Moose droppings!

Most mosses produce small, rather inconspicuous sporophytes-stalks bearing a capsule full of dustlike spores. The spores are shaken out of the capsule like salt out of a salt shaker and then blown through the forest on the wind. That works fine if the moss isn't too choosey about where it grows.

***Splachnum* has come up with another adaptation that helps ensure that its spores land on Moose dung. Its capsules are expanded outwards like small parasols, which are attractive landing pads for flies attracted to the Moose dung. *Splachnum* also has sticky spores that get stuck on the feet of the flies and are transported to the next pile of droppings. A closely related genus, *Tetraplodon,* lacks the parasol landing pad but does have the sticky spores. *Tetraplodon* is usually found on carnivore dung or old carcasses of small mammals.**

Splachnum

its alternative name, mountain misery. False Azalea has the loose whorls of leaves of the rhododendron-azalea group, but its flowers are tiny and bell shaped, like those of blueberries, and are an unusual copper colour.

The floor of dry, open, mountain forests is carpeted with the delicate green foliage of Grouseberry, Mountain Arnica and Western Meadowrue. In wetter areas, Oak Fern, Knight's Plume Moss, leafy mosses, Queen's Cup, Sitka Valerian and Rosy and Clasping Twistedstalks predominate.

Oak Fern

MUSKEG FORESTS OF THE NORTH

Trees

Black Spruce

Tamarack

Black Spruce always looks like the poor cousin of the spruce family. This is the scrawny conifer of northern muskegs, generally growing thin and straight, with a clubbed top. It is found as far south as Mount Robson in the Rockies and Quesnel on the Fraser Plateau. Although it is generally associated with peatlands, in the north it often grows on well-drained areas, as Lodgepole Pine does. Black Spruce has small, tough cones that squirrels find difficult to deal with. In addition to producing seed-bearing cones, this species can also reproduce by layering. As the lower, live branches become covered by moss, they put down roots and become trees of their own.

Living with Black Spruce in the northern wetlands is Tamarack. Tamaracks, like other larches, are deciduous conifers; they lose their needles each fall. In the spring new, soft, light green needles reappear, adding a delicate touch to the muskegs. Despite living in cool, wet places, Tamaracks demand sunshine and do not grow in shady woods.

Biting Flies

Nothing links humans with the insect world more effectively than a cloud of black flies—except perhaps a single, whining mosquito in a bedroom at the first light of dawn, or maybe a gigantic horse fly circling a canoe. Our link with these flies is their desire to sip on our blood. Female black flies, mosquitoes, horse flies and biting midges all require blood meals to maximize their egg production; males do not need blood and are not attracted to humans.

After mating, the female mosquito begins her search for a blood meal, a search that may take her a number of kilometres from her birthplace. Carbon dioxide is the first clue she seeks, since all animals exhale it in quantity. Flying towards increasing concentrations of carbon dioxide, she then detects heat at relatively close range if the victim is warm-blooded, and finally she sees you. The most effective mosquito repellent, DEET, interferes with her detection of body heat.

After doubling or even tripling her own weight with a blood meal, the female retires to a sheltered place to allow her eggs to develop. When she has laid her eggs, she immediately seeks out another blood meal—and the cycle continues. In southern British Columbia, mosquitoes may be able to complete up to five egg-laying cycles in a season.

The larval stages of all these biting flies are associated with water. Mosquitoes prefer standing water, and depending on the species (there are forty-five in British Columbia), larval "wrigglers" can be found in a variety of ponds, marshes, bogs and even tree holes.

Black flies are smaller but chunkier than mosquitoes; they are usually grey or black, but one common

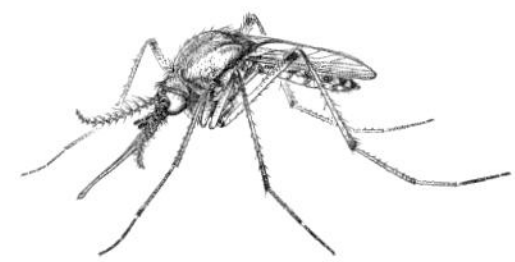

Mosquito

Black fly

Horse fly

Pages 62–63: **Bunchberry blankets in the dark forest blaze pure white in June and produce red berries in August.**

group is bright orange. After they land, they scurry around looking for a good place to feed and usually settle on areas with thin skin that are somewhat protected, such as under a collar or at the base of a bushy beard. Because black flies use a "slash and lap" method of feeding and employ an effective anticoagulant, their bites are often very messy events, with blood flowing freely.

Opposite page: **This Common Raven clearly shows the heavy bill and wedge-shaped tail that distinguishes it from its crow cousins.**

Larval black flies develop in running water, explaining why a cloud of adult females may descend on you as you take a hiking break beside a clear mountain stream. In the stream the larvae glue themselves to rocks and use large brushes on their heads to filter out the tiny plankton rushing by.

Ravens

One of the symbols of the northern forests, the Common Raven is a carrion eater. Territorial adult ravens habitually follow wolves and other predators to claim the leftovers at a kill site, sharing the carcass with only their mate. If a single, young raven finds a dead moose or similar choice feast, however, it gives a loud, distinctive call known as the "yell," which tells every raven within earshot that dinner is available. Soon a large crowd of ravens, mostly young birds, gathers at the carcass to share the meal.

Why are young ravens so happy to share while adults quietly gobble all they can? It's simply a matter of strength in numbers. If a young bird tries to hog a dead moose for itself, the local territorial adults will soon find out and unceremoniously chase the youngster off. But if the young bird calls in all the local unattached ravens, they can fend off the territorial pair and share a meal that could well allow them to survive a few more days in the cold northern winter.

Many ravens in British Columbia have discovered a new food source to help them through the winter—garbage. Most landfill sites across the province are a Mecca for ravens in winter; as a result, more young birds make it through the winter and the raven population is soaring.

The tiny flies known to many of us as no-see-ums are biting ceratopogonid midges. The adults are large enough to see but small enough to scramble through regular mosquito netting. Their bites are fiery and out of all proportion to the biter's size. The larvae are small, clear worms that live in a variety of aquatic habitats.

Horse flies and deer flies are at the large end of the scale of biting flies. These two closely related groups can be distinguished in several ways. Horse flies are larger, have clear wings and usually have iridescent stripes through their eyes. Deer flies are smaller, have dark marks on their wings and have iridescent spots on their eyes. Unlike mosquitoes, which prefer the warm, humid early morning and evening hours, horse flies and deer flies thrive in the hot sunshine of midday.

***Opposite page:* Young Subalpine Firs and Engelmann Spruces grow among the straight trunks of a Lodgepole Pine fire pioneer forest.**

FORESTS OF FIRE

Drifting in on a warm evening breeze, the sweet, soft smell of a forest fire is one of the scents of summer in the Interior of British Columbia. Spectacular summer thunderstorms are common, and the lightning bolts flashing out of the sky start thousands of fires each year. Most of these conflagrations remain small, but some grow to a considerable size, burning thousands of hectares; all do their part to create a complex and dynamic forest. In fact, old-growth spruce forest in the Interior can be scattered and rare, and the fire succession forests of Lodgepole Pine and Trembling Aspen are much more common.

In the early part of this century, fires swept unchecked through the cold forests of British Columbia at an unnaturally high rate—fires accidentally set along new railway lines or started by prospectors to find patches of bedrock hidden by the forest. Vast tracts of old spruce and fir were vaporized and replaced by young Lodgepole Pine forests. Although most of these pine forests would eventually grow back over the centuries into spruce and fir, modern logging practices have changed the natural order of succession. And although the old spruce forests and the young forests of fire are adapted to the same harsh winters, they are very different forests.

Lodgepole Pine

Lodgepole Pine is well equipped to take advantage of a fire, since many of its cones are sealed with resin, remaining tightly closed until they are opened by intense heat. After the fire roars through the forest, a rain of tiny, spinning pine seeds floats to the ground, which has just been conveniently fertilized with ash and cleared of brush, grass and other competitors. If a fire doesn't occur, the sealed cones remain on the tree's branches, and some are eventually enveloped by the bark and wood of the growing tree. The seeds in these imbedded, sealed cones can remain viable for 150 years or more. Lodgepole Pines also produce cones that open as soon as they mature, so the species can also colonize open areas where fire has not occurred.

In the bright sunlight of the fire meadows, the young pines grow quickly. If too many seedlings grow up, however, they form an amazingly thick forest of straight, skinny trees—called a dog's hair forest—and their growth can be slowed and stunted. Sometimes trees a little over a metre tall can be seventy years old. Over the years, most of these trees are shaded out by faster-growing individuals and die and fall, creating a tangle resembling a pick-up-sticks game. The surviving pines are tall and slim—they can be 15 metres tall but only 15 centimetres in diameter! Between the slender logs, spruce and fir seedlings grow quietly in the cool, humid shade and after a couple of hundred years overtop and replace the sun-loving pines.

In the dry, western Chilcotin Plateau, in the rain shadow of the Coast Mountains, only a minor amount of spruce regeneration occurs, and older Lodgepole Pines form open, sunny forests. The forest floor there is blanketed with the shiny evergreen leaves and bright red berries of Kinnikinnick, a soft cushion of haircap mosses and feather mosses, and a lush mat of lichens, particularly Reindeer Lichen. In some areas, Soopolallie bushes provide abundant fall food for bears and other berry-loving animals.

Lodgepole Pine

Lodgepole Pine has two common parasites, which often catch the eye of naturalists. The most obvious are the large galls found on the trunks of young pines or the branches of older trees. Like the witches' broom on spruce, these are caused by a rust fungus, the Western Gall Rust. The galls take two to four years to mature, after which orange spores appear on the surface of the gall. The spores are blown by the wind through the forest and, unlike most other rusts, which require an alternate host, directly infect other pines. Western Gall Rust rarely kills the pine, but galls on the trunk are a major source of weakness and the trunk often snaps at this point before the tree reaches maturity.

A less obvious parasite, but one with a greater impact on the health and growth of the pine, is the Western Dwarf Mistletoe, a tiny, yellow-green leafless plant growing on the pine branches. The mistletoe roots burrow into the branch to steal sugars, water and other

A dwarf mistletoe parasitizing Lodgepole Pine.

essential nutrients from the host tree, reducing its vigour, but they usually do not kill the tree. Mistletoes are flowering plants that produce berries with sticky seeds. Dwarf mistletoes disperse their seeds in an unusual way. When the single-seeded "berries" are mature they drop from the stalk, and at the same instant the skin around the seed splits open and contracts violently. This action propels the seed at speeds of up to 80 kilometres an hour, much as a moist orange seed or cherry pit will be propelled if squeezed between the fingertips. In this manner seeds can infect branches up to 10 metres away. You can see that mistletoe can spread much more easily in young, dense stands of pine than in thinned stands where trees are much farther apart.

One of the most important forces directing the logging industry on the plateaus of central British Columbia for the last fifteen years has not been catastrophic fires or the near extinction of old-growth forest but a tiny insect named *Dendroctonus ponderosae*, the Mountain Pine Beetle. Although its scientific name suggests that it enjoys Ponderosa Pines as well, this beetle really thrives on the vast monocultures of Lodgepole Pine carpeting the province's Interior.

Mountain Pine Beetle

Adult beetles go on the attack in July and August, searching out mature pines, preferably greater than 25 centimetres in diameter. They bore through the thin outer bark into the nutritious cambium beneath. The tree immediately fights back by exuding copious pitch, but only the most vigorous of trees growing in ideal conditions can withstand a concerted attack by many beetles. Infested trees show pitchy holes by late summer and reddening foliage by spring. As the beetles chew out egg chambers in the cambium, they bring with them spores of Bluestain Fungi, and the combination of voracious beetle larvae and fungal infection girdles and kills the tree. Next summer, the larvae mature and emerge as adults, which fly off to look for more suitable trees. In the British Columbia Interior such trees are rarely far away.

When they begin feeding in a tree, the beetles release pheromones, simple volatile chemicals that attract other beetles, and so another pocket of infestation begins or a small infestation becomes larger. When the tree becomes crowded with beetles, the males emit another pheromone, which deters other beetles from landing.

Serious epidemics usually begin during drought periods, when trees are stressed and summer temperatures are ideal for larval growth, and end a decade or so later when a cycle of higher precipitation begins. In the middle of the dry 1930s, two biologists working for the Royal British Columbia Provincial Museum reported: "Throughout the Chilcotin we found that bark-beetles had attacked the Pine to an alarming extent. The hillsides were covered with vast patches of the dying trees." The forests recovered over the next few decades, and then a series of dry years in the 1970s triggered another major outbreak. When we drove through the forests of the western Chilcotin in 1978, the scene was very similar to that described in 1935. This epidemic spread throughout the southern province, peaking in 1984, when more than 400 000 hectares of pines were infested.

Large pine beetle infestations speed up the succession processes in most forests. Young firs and spruces, released from the shade of the pines, grow quickly into a climax forest. In dry forests where Lodgepole Pine is the climax species, such as forests in parts of the western Chilcotin, infected stands quickly produce more mature Lodgepole Pines for the beetles to attack, and so epidemics can be more damaging and long lasting. Outbreaks can enhance the diversity of these forests, though, by creating a mosaic of unevenly aged stands.

The forest industry, however, is unwilling to wait for the fir and spruce forests promised by widespread beetle attack. Immediate harvest has been the usual way of dealing with pine beetle infestations, resulting in the largest clearcuts in the world in parts of the Fraser Plateau. Unfortunately, these large clearcuts usually grow back into monocultures of Lodgepole Pine, gunpowder for a new beetle explosion.

New techniques using pheromones to capture or at least divert beetles searching for new trees are effective on a small and short-term scale, but as long as huge areas of the province are covered with dense stands of Lodgepole Pine the Mountain Pine Beetle will continue to flourish. Silviculture practices that produce a great diversity of tree species and ages are the best way to reduce pine beetle epidemics. Careful thinning of uninfected pine stands, for instance, speeds the growth of spruce and fir seedlings, produces vigorous pines and increases wind flow to disperse beetle pheromone trails.

Pages 74–75: **In the fall, patches of genetically identical aspens can easily be picked out from the rest of the forest, since the trees within them all turn colour at the same time.**

Aspens

In the autumn, the northern forests are a patchwork of dark green conifers and blazing yellow and orange Trembling Aspens. As its flickering, golden leaves seem to signify, the aspen is another tree of fire. Like its poplar relatives, aspen is a fast-growing, short-lived tree; few individuals are more than a hundred years old. Like most western deciduous trees, aspens need full sun to grow. The fluffy, buoyant seeds of aspen and other poplars are unusual in that they must germinate within a few weeks of hitting the ground. They cannot lie dormant as other plant seeds do, waiting for proper moisture levels. Seedling aspens are therefore rare under normal conditions. A stormy summer produces just the germinating conditions aspens need—lightning fires open sections of the dark forest and occasional downpours water the seeds.

Trembling Aspen

Like its other poplar relatives, however, the aspen usually propagates through suckers, and following

disturbances such as fires it can reproduce in abundance in this way. Often extensive clones form, each originating from a single parent tree.

Aspens are valuable wildlife trees for several reasons. Moose and Elk browse their twigs and bark in winter, and aspens growing close to a pond are a favourite with Beavers. But aspens are especially favoured by woodpeckers. Most woodpeckers usually feed outside the aspen woodland—Pileated Woodpeckers in mature coniferous forests, flickers looking for ants on the ground in meadows, sapsuckers in willow and birch thickets—but they prefer to nest in aspens. Why? Most mature aspens are infected with heartrot fungus, and woodpeckers find that excavating a nest hole in an aspen is simply a lot less work than chipping away at some other tree.

Next time you are in an older aspen grove, look around and count the number of woodpecker cavities—several can usually be found in a few minutes. These homes are now available for squirrels, chipmunks, mice, small owls, swallows, bluebirds, wrens, chickadees, nuthatches, honeybees, bumblebees and the many other animals that use tree cavities for roosting and rearing young.

Another inhabitant of aspens, the Forest Tent Caterpillar, can have a dramatic effect on these forests. In 1974 I (Richard) was working in the Rocky Mountains of British Columbia. I arrived in late May as spring was in full sway, the aspens a beautiful pale green against the dark spruces and firs. But within a matter of weeks the beautiful wilderness scene quickly changed to one of apparent ecological disaster—every aspen leaf in the valley had been devoured by a swarm of millions of

Forest Tent Caterpillars. The caterpillars, rather attractive on close inspection—black with blue and orange spots—were literally raining from the trees.

Local residents told me that tent caterpillars had devoured the aspen leaves the year before as well. I was amazed that the aspens had survived, but I soon found out how they did it. By July, the woods looked like early spring again—a flush of light green came over the white aspen branches as new leaves sprouted from the axils of the old leaf stems. The fat, mature caterpillars wove their small silken cocoons, many of which became food for chickadees and warblers. Those that survived soon emerged as small brown moths, equally attractive food items for warblers and vireos.

The following year, there were almost no Forest Tent Caterpillars in the Robson valley. What had happened? Tent caterpillars, such as the Douglas-fir Tussock Moth and the Spruce Budworm, have cyclic populations. Every eight to eleven years, tent caterpillar numbers increase to astronomic proportions for a couple of summers and then disappear over the next few years before mushrooming again a decade later.

In the year of the decline, the caterpillars first appear healthy but soon become lethargic and then begin to die by the thousands. The cause of death, and perhaps the driving force of the entire cycle, is a disease agent called the nuclear polyhedrosis virus. Viral bodies on leaves are eaten by the caterpillar and replicate in the nuclei of the caterpillar's cells. Within a week or so, millions of viral bodies are produced and the caterpillar bursts, leaving a limp cadaver on the leaf and a huge number of viral bodies to spread the disease to other caterpillars. When most of the caterpillars are dead, the

viral bodies have very few hosts in which to replicate, and outside their hosts, the viral bodies are slowly destroyed by ultraviolet light. A few viral bodies survive in shaded areas and after ten years or so begin to build up in the caterpillar population again as the moths start to multiply. And so the cycle continues.

WINTER IN THE SPRUCE KINGDOM

Opposite page: A tachinid fly has laid its white eggs on this tent caterpillar; the eggs will hatch into maggots, which will eat the larvae from the inside.

The primary story of these great forests of spruce, fir and pine is the story of winter. To fully understand them, naturalists cannot simply visit them at the height of summer when the meadows are in full bloom, tropical migrant birds sing in the trees, and insects swarm in the air. Winter is the time of reckoning here, and a midwinter trip to the spruce forests reveals many of the factors that shape them.

Animals of Winter

Winter is a quiet time in the mountain forests. Often the only sounds heard over the trees creaking in the cold wind and the trickling of ice-bound brooks are the croaks of a lonely raven and the occasional twitters of a roving band of chickadees and kinglets. Gone are the buzzing insects and singing birds of the previous summer. The Elk have finished their autumn bugling, and the Bighorn rams have stopped bashing heads for another year. Winter is serious business here, and any animals that are active are working full-time to survive.

Three-quarters of the birds have moved out. Half of the bird species, mostly insect eaters such as flycatchers, swallows, vireos and warblers, have gone to Mexico and beyond. Another quarter have made more modest journeys, some to California, some to the gardens of the

Lower Mainland, others to the more temperate valleys of the southern Interior. The quarter that stay in the boreal forests, about twenty-eight species in all, belong to one of four loose categories: predators (hawks and owls), bark gleaners (woodpeckers, chickadees, nuthatches), scavengers (jays and ravens) and seed and bud eaters (grouse and finches).

Many of these winter birds will also move out when food is scarce. Owls are forced south to find alternative prey if hare or vole populations have crashed, and if local forests are insect free, Hairy Woodpeckers will move into southern forests, searching for beetle infestations. For some species, availability of food is so unpredictable that wandering becomes a way of life; two rather bizarre species of finch, the crossbills, exemplify this nomadic lifestyle.

Crossbills are unique among the birds of British Columbia—the tips of their bills are crossed like the points of fine scissors. This feature, which looks as if it would hinder attempts to feed on anything, is an adaptation for extracting seeds from cones. In fact, crossbills rely on conifer seeds so heavily that they roam the continent in search of areas with good cone crops. One year they might be the commonest bird species in British Columbia forests, and the next almost all of them are gone, feeding on spruce seeds in New Brunswick or pine seeds in Oregon. Where cone crops are heavy, crossbills can breed year-round; we have seen courting and nest-building pairs during our Christmas Bird Counts.

White-winged Crossbills are the smaller of the two species and specialize in spruce cones, so they are restricted to the boreal and subalpine forests of British

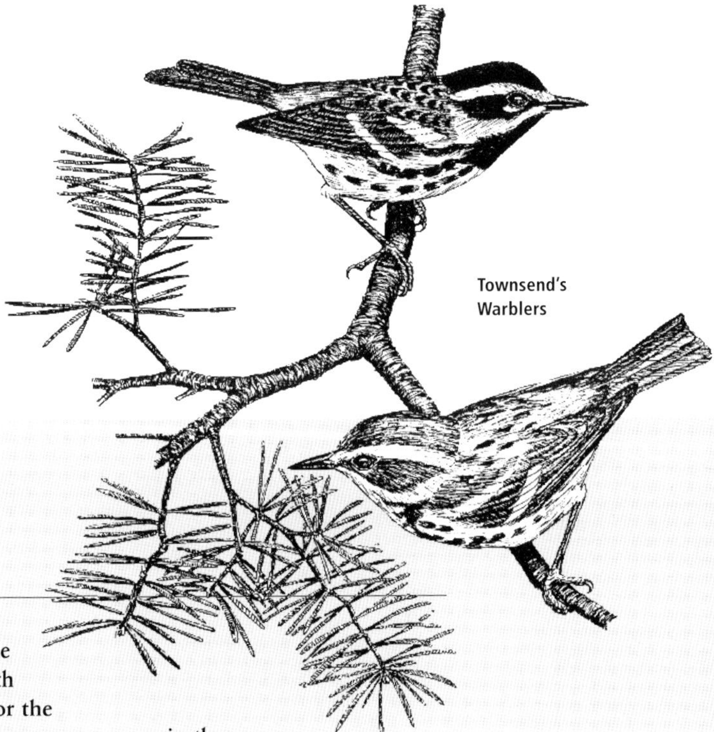

Townsend's Warblers

Fitting In

Most of the birds that escape the cold winters of North America each year for the balmy tropics don't go as far as you might think. Most stop in southern Mexico and Guatemala, and a few make it to Costa Rica—the tropical forests of South America are the major destination of only about a half-dozen of British Columbia's species. But Central America is only a fraction of the size of North America. How do most of the North American migrants fit into that small area, which is already teeming with birdlife?

The simple answer is that they haven't been gone long. Most of the birds that are with us in the summer—swallows, flycatchers, warblers and others—are really tropical species that dash north to breed. They take advantage of our long summer days, hordes of insects and lack of snakes to produce as many young as possible and then wing their way back home to Mazatlán and Monteverde. Their stay-at-home jungle neighbours simply don't have enough time to increase their numbers and take over the gaps vacated by the northern travellers.

Below: A male Red Crossbill searches for seeds in Ponderosa Pine cones.

Opposite page: A blanket of winter snow insulates trees from winter winds.

Columbia. Red Crossbills feed on a wider variety of conifer seeds, and although their populations seem more predictable on the surface, recent research has revealed that there may be as many as six species of "Red" Crossbill in British Columbia, each specializing in a particular conifer species.

So why are their bills crossed? Craig Benkman of the University of British Columbia has shown that this unusual mechanism is needed to extract seeds from

unopened, immature cones, an essential ability if crossbills are to eat conifer seeds year-round. Crossbills insert their closed bills between the scales of a cone and then pry the scales apart by moving the lower mandible sideways. When the gap is large enough, they extract the seed with the tongue.

Many of the birds of the boreal winter find enough food simply because they have personal storehouses. Chickadees wouldn't survive the long, cold winter if they didn't store insects, spiders and seeds in tree cracks all summer. Did you ever wonder what Gray Jays eat when there are no hikers or skiers around? The next time a pack of these jays (also known as whiskey-jacks or camp robbers) discovers your campsite or picnic lunch, watch them—most of the time they don't eat the food right away but fly away with it to store it in a secure place.

Small birds such as chickadees, nuthatches, creepers and especially the tiny Golden-crowned Kinglets, are faced with a life-threatening situation every winter

Boom, Bust and Owls

Boreal Owls, like many hawks and owls, specialize in eating meadow mice, or voles. These mice have four-year population cycles, similar to those of lemmings.

Recent research has revealed that Boreal Owls have some remarkable breeding strategies. Older Boreal Owls, which have gone through one boom-and-bust cycle of mouse numbers, put a tremendous effort into raising young the first year that mouse numbers begin to climb, even though the rodents are still rather uncommon then. The following year, when mice are everywhere, the older owls relax and raise fewer young. Younger, less experienced owls raise large broods at the peak of the mouse cycle, only to have their young starve the following winter as mouse numbers crash.

night—freezing to death. Their small size forces them to use a tremendous amount of energy staying warm. Kinglets (which weigh only 6 grams) manage to keep their body temperature at 39° to 42°C all winter, even in -30°C weather. They feed all the short winter day long to get enough energy to survive the long winter night but would still die on a cold night if they didn't roost communally. Research has shown that kinglets reduce heat loss by 23 per cent if they roost in pairs and by 37 per cent if they sleep in a ménage à trois. So at

This male Three-toed Woodpecker (identified by its yellow crown) flakes the outer bark off a spruce trunk in a never-ending search for bark beetles.

Three-toed Woodpeckers

If you are out skiing or snowshoeing through spruce forests in midwinter, look for woodpecker sign. Listen for quiet tapping and look for reddish trunks where Three-toed Woodpeckers have flicked away the grey scales of spruce bark in their search for beetles just underneath. Often the red bark of infested trees stands out from a great distance, and bark scales litter the snow. Why this species has lost one of its toes—most woodpeckers have four on each foot—is still a mystery.

A male Spruce Grouse struts for a nearby female, fanning its tail to show off the characteristic patterning of the mountain form—often called the Franklin's Grouse—of the species.

the end of the day, a kinglet's number one priority is to find a sheltered spot and some kinglet friends to snuggle up with and conserve energy. That is why you never see a kinglet by itself on a winter day.

The characteristic grouse of these spruce forests is, not surprisingly, the Spruce Grouse. Blue Grouse hoot

from the mountain meadows and Ruffed Grouse drum from deciduous draws, but Spruce Grouse have the thick forests more or less to themselves. Like the Blue Grouse, the Spruce Grouse has two distinct forms in British Columbia. One is found east of the Rocky Mountains and across the northern edge of the province, and another is found on the central plateaus and mountains. How are they different? Types of grouse almost always differ in the tail, since the tail is the centrepiece of male display in most chickenlike birds, from peacocks to turkeys to Spruce Grouse. All male Spruce Grouse have a jet-black tail; in the eastern form it is tipped with a reddish-brown band. The mountain form—often considered a separate species, the Franklin's Grouse—has an entirely black tail, but the tail coverts are tipped with white, producing a spectacular spangled effect when the tail is fanned in courtship.

Pages 90–91: **Golden-crowned kinglets are among the most common birds in northern and mountain forests, yet they are rarely seen as they forage in the treetops.**

There are differences in the courtship itself as well. Eastern Spruce Grouse males attract females with a loud, fluttering flight up to a favourite branch. Franklin's Grouse do the same but add a distinctive touch to the descending flight. Halfway to the ground, the male throws his wings over his back and cracks them together twice, producing a loud *clap, clap,* sounding just like some very short but enthusiastic applause in the quiet of a spring dawn in the mountains.

Spruce Grouse spend the winter in a quiet, retiring manner up in the trees, patiently eating the fresh needles produced during the last summer. Grouse will often pick the same tree to feed in year after year—studies in the Yukon show that these favoured trees are among the most productive spruce in the vicinity, despite the continuing depredations of grouse. The

needles of these trees are apparently more nutritious than those of other trees, and the piles of grouse droppings that build up underneath them help keep them nutritious. Look for large trees with lots of needles missing near the branch tips.

Although a few mammals of the boreal forest hibernate and others live on stockpiles of food, most survive the winter just by getting by—remaining active, continually searching for food and always losing weight. For large herbivores such as Moose, times are hard because the nutritious green food they lived on all summer is gone; the plants are covered by snow or ice or have sent many of their nutrients into their root systems. Moose rely almost totally on shrubs and trees at this time, browsing the relatively nutritious twig tips from the previous season's growth. To reduce metabolic costs in winter, they lower their body temperatures somewhat, but despite this strategy their food intake is about 30 per cent less than that needed to maintain their body weight.

When you are out on your summer hikes, look for signs of browsing—willows and smaller shrubs often are obviously well pruned. Since Moose don't look under the snow for food, conifers sometimes have a skirt of snow-protected branches below an empty break; the height of the skirt tells you the depth of the winter snow, and the height of the upper branches shows how high the long-necked Moose can reach. By looking closely at the clipping points on the old branch tips, you can guess what animal did the browsing—Snowshoe Hares make neat diagonal clips, whereas Moose and other deer, lacking upper teeth, tear off the branches, leaving messy, frayed ends.

Thus, winter is a time of weight loss for most animals—and malnutrition and frigid weather make individuals susceptible to many diseases. It is vitally important for animals such as Moose to enter winter healthy and fat. Many deer and sheep mate in the fall, however, and the males don't have time to eat much when they are busy fighting other males and courting females. Consequently, it is the older, reproductively active males that are often most likely to die during winter.

Everyone knows that squirrels gather and store food for the lean winter ahead. In British Columbia forests, look for the middens of Red and Douglas' Squirrels—great heaps of cones lying under conifer trees, with tunnels leading into the core of the larder. Squirrels also gather mushrooms—look for their drying racks up in the branches of trees.

Squirrels aren't the only animals that prepare for winter by storing food. Pikas, for instance, gather grasses and forbs at the height of summer and make hay piles for curing their winter food, and Beavers gather branches all summer and store them in their deep ponds for winter chewing.

In winter, most insects become inactive and spend the cold, dark months in some sort of hibernation. A few, however, are most conspicuous during cold, snowy weather, even in the high mountains or the far north of the province. In British Columbia, the two groups of snow-loving insects most commonly encountered are the wingless snow flies *Chionea* and the snow scorpionflies *Boreus*.

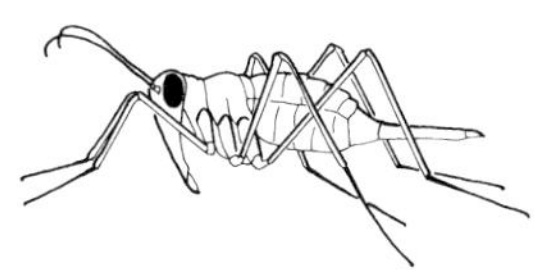

Snow scorpionfly

Snow flies are in the crane fly family but are much smaller than the garden-variety crane flies, and their wings are reduced to stubs. With slender but strong,

arched legs, they look like slim spiders as they scramble over the snow (count their legs, though—there are only six!). The name *Chionea* comes from the Greek *chion*, meaning snow.

In the summer the larvae live in mammal burrows and forest litter, but little is known of their way of life. Experts speculate that unlike most of their vegetarian crane fly cousins, snow fly larvae are predators, feeding on various other small invertebrates. Adults seem to spend most of their lives in mammal burrows, under rocks or logs, or in the forest litter in the narrow air-space beneath the snow. Under the snow they are well insulated from the cold, for sub-snow temperatures remain quite constant and close to freezing even when

Opposite page: **The Porcupine's winter diet of tree bark is so low in nutrition that the animals lose weight all winter even though their stomachs are full.**

Porcupines

Porcupines are a unique member of the northern forest fauna, the lone Canadian representative of a family otherwise restricted to more tropical climates. Porcupines left their South American homeland about two million years ago when that continent bumped into what is now Panama.

A Porcupine's formidable quills are a deterrent to most predators but have some built-in safety features in case the Porcupine accidentally punctures itself in a fall. For one, they contain a mild antibiotic so that the wound, though painful, rarely becomes infected.

Like many other animals, Porcupines are strict vegetarians, eating leaves in summer and the inner bark of trees in winter. Because this diet is very poor in sodium, Porcupines (and other grazing animals) are strongly attracted to any salty items they happen to find. Outhouses and signs don't often last for many years in British Columbia forests, since Porcupines have developed a fondness for the salty glue in plywood. Foods that are high in organic acids upset this sodium balance further, and so porcupines search out favourite trees that are blander in taste than other trees in the forest.

the snow cover is relatively thin. The adults usually appear on the surface of the snow only when the temperature is hovering around freezing and when the light is dim, such as at dusk or on a cloudy day. At these times, the surface temperature is relatively stable; if the temperature dropped suddenly, even these winter insects would freeze and die.

Opposite page: **Great Gray Owls hunt by sound, using their huge, dinner-plate faces to gather in the tiny noises of scurrying mice and even burrowing pocket gophers.**

Snow scorpionflies in the genus *Boreus* are small, glossy brown or black insects that resemble minute grasshoppers with long snouts. The males have tiny, bristle-like wings with which they hold the females during mating; the wings of the females are reduced to just a pair of small scales.

Away from snow, *Boreus* are rarely seen and are invariably associated with mosses, especially those that grow in low, compact cushions. It is curious that these snow-loving, moss-eating insects are British Columbia's only representatives of the predominantly tropical, predatory scorpionflies.

Living with Snow

Snow can be a problem or a boon to big animals. Early in the winter, big falls of powder snow can be advantageous to animals such as Moose, for their long legs can travel through it with relative ease, whereas predators such as wolves may flounder. As the snow develops a strong crust, however, the advantage swings over to the wolves, who can run along the top of the snow at speed, whereas the Moose continually break through it. To combat this problem, Caribou have specially adapted hooves that spread out and essentially form small snowshoes—the ratio of their foot area to their weight (a foot-load index) is more than twice that of Moose.

Caribou have broad hooves that double as snowshoes and snow shovels—in shallow snow they are invaluable in clearing snow craters to reach food.

Caribou use their snowshoe hooves to move into forests high on mountain slopes, places that wolves shun since the snow there is deeper and stays soft much longer than in the valleys. By travelling in herds, Caribou can also share the task of moving through deep snow by taking turns leading and plowing the trail.

Mice and other small mammals live under the snow, which, like a thermal blanket, protects them against the severe cold above, though it may make food harder to find. Even in subalpine and alpine regions, temperatures in Deer Mice burrows hardly ever fall below freezing.

The commonest mice of the spruce forests are the Southern and Northern Red-backed Voles, which subsist on seeds, fungi and lichens. They peak in population in years of bumper spruce cone crops, and at that time, Boreal Owls and Northern Saw-whet Owls congregate to feed on them. Hawks have a harder time finding mice hidden under snow, so most head to the southern coast and valleys in winter. But owls stay behind and put their special talent—hearing—to work. The Great Gray Owl can hear and catch mice under a half-metre of snow.

The Northern Red-backed Vole is probably the commonest small mammal in the spruce forests of British Columbia and is therefore a favourite food of the Boreal Owl.

PREDATOR AND PREY: CONNECTIONS IN THE SPRUCE FOREST

The Snowshoe Hare Story

We first ventured into the northern forests of British Columbia in the summers of 1979 and 1980. Of our many memories of that summer, we most clearly remember the long days, the long roads and the abundance of animals—Snowshoe Hares almost constantly in view, bounding into the forest; dozens of lumbering Porcupines; a Lynx padding off into the aspens; families of just-fledged Great Horned Owls perched on roadside spruces in the long twilight; and, in the subalpine willows, ptarmigan literally everywhere. We returned to the same roads only three years later but found them almost empty of wildlife—we saw no Porcupine, no Lynx, one pair of owls, only one or two hares and a small flock of ptarmigan only after a strenuous hike into the mountains.

Opposite page: **Lynx have long legs and huge feet to carry them long distances over the snow, searching for Snowshoe Hares.**

This boom-and-bust pattern in the populations of northern animals has long intrigued ecologists. Most of the species that exhibit this pattern show clear eight- to eleven-year cycles, and several apparently unconnected species show some synchrony in their cycles. The key to this puzzle—and one of the keystone species in the ecology of the entire boreal forest—is the Snowshoe Hare.

During the low portion of the population cycle, hares can be hard to find, with only about ten animals per square kilometre. This low phase lasts four or five years, and then the population begins to climb rapidly for two or three years, peaking at densities about fifty times that found in the low phase. Hares are literally everywhere at this point, the normally nocturnal animals feeding brazenly by roadsides during the day. Predators are everywhere as well, especially the hare specialists—

Snowshoe Hares are familiar to most naturalists as the "rabbit" of the northern forests that turns white in winter. Basking in the April sun, this one is already showing signs of its brown summer coat.

Lynx, Great Horned Owls and Northern Goshawks. Other species also thrive during this period, perhaps taking advantage of the fact that most of the predators are concentrating on hares and temporarily disregarding burgeoning populations of grouse and ptarmigan. The large number of hares is usually maintained for two years, and then in the winter of the third year 95 per cent of the hares die and the population is at the low level again.

The two primary theories of the cause of the Snowshoe Hare cycle are intuitively obvious: (1) they eat themselves out of house and home, and eventually most starve to death, and (2) their predators are able to increase their own populations quickly and simply eat most of the hares. Or the cycle may be the result of a combination of the two factors. Several groups of ecologists have worked extensively on this problem, since it seems to be the key to understanding the ecological processes in the entire boreal forest. One group of scientists from the University of British Columbia and several other Canadian universities has been studying hares and other animals in the southwestern Yukon

Porcupine tracks

Tracks in the Snow

Skiing or snowshoeing across a fresh blanket of snow can be a revelation for the curious naturalist—everywhere there are signs of life in the seemingly lifeless winter surroundings. Tracks of animals that are so secretive that they are rarely encountered face to face are there for all to see, and the stories of nocturnal encounters of predator and prey are spelled out on the snow.

For this reason, biologists often use tracks to study the lives of animals they have difficulty studying directly. Tracks can be used to answer questions such as: "What predators live in this forest? How common are grouse? How far does a wolverine travel? Where do lynx go? What do they eat?" If you wait a particular time after the last snowfall—say, two days—you can use tracks as an index of the abundance of animals from time to time and from forest to forest.

Whose tracks are these? A Porcupine dragged its tail through fresh snow as it waddled by.

since 1977 and is beginning to shed light on this puzzle. In areas where hares were fed extra food and protected from predators to some extent, high numbers persisted as the population crashed all around; populations given extra food but not protected crashed as well. So it seems

Avalanche tracks in Kootenay National Park.

Avalanches

Avalanches are one of the most obvious examples of how snowfall shapes mountain forests. Vertical, grass-green swaths through the dark conifers on steep slopes are the tracks of winter and spring snowslides. Avalanche tracks are dominated by deciduous shrubs, especially alder, since these plants can regrow from sprouts if the main stem is snapped off by the catastrophic power of an avalanche. If a conifer trunk is broken off at the ground, the tree dies. Since most of the snow on avalanche tracks is swept away to the valley bottom, these areas are often the first places where green grass and herbs appear in the spring. Because these plants attract hungry herbivores waking from their winter sleeps, bears, ground squirrels and marmots are often seen on avalanche slopes in early spring.

that predators play a key role in reducing the hare population, and this role is probably enhanced by the reduction in food supply.

The Yukon study has also provided surprising new data supporting an old but often challenged theory—that there is a connection between hare numbers and sunspots. The hare cycle can be detected for two hundred years or so in the past by examining cores from trees and counting the characteristic marks left by browsing hares in each year. Climatologists have cored even more ancient structures—the glaciers on the nearby St. Elias ice fields—and have discovered a ten-year cycle in precipitation. This cycle not only matches that of the hare cycle but lines up surprisingly well with records of sunspot activity kept for hundreds of years by European astronomers. Most intriguing of all, cycles in which sunspot activity changed only moderately fell out of phase with the hare cycles, whereas strong sunspot cycles matched those of the hares exactly. Could sunspot activity affect rainfall patterns, which in turn could affect plant growth, which could boost the hare population to unusually high levels?

Connections between Caribou, Moose and Wolves

The web of life connects the inhabitants of the mountain forests in many ways, but there are few more dramatic connections than those between big predator and big prey. Moose were rare in central British Columbia at the turn of the century. Wolves hunted small herds of Caribou and what Moose they could find before the fires and logging of this century produced vast areas of prime Moose habitat. As the Moose population soared, the wolf population climbed as well.

Lovers of willow thickets created by wildfires and logging, Moose have prospered in the last few decades.

So who are the losers in this race? The Caribou, of course. Wolf populations are now much higher than they have been for years, and although they dine on Moose, they continue to take Caribou whenever it is

convenient. The Caribou now suffer more wolf predation than ever before, and the same fires and logging that benefitted Moose have shrunk the old-growth mountain forests, the critical winter habitat of Caribou. Although Caribou hunting has been curtailed, many herds around the province have greatly declined.

The problem facing wildlife managers is whether or not to try to help solve the problem using predator control. Wolf kills have galvanized protests by environmental groups, and some biologists question their effectiveness. Moose, the unknowing wild card in this hand, continue to support wolf populations.

THE FUTURE

The mountain forests and meadows of British Columbia, largely because of their spectacular beauty, are well represented in the provincial parks system and are not considered endangered ecosystems. In comparison with habitats restricted to southern lowlands, such as the grasslands of the south Okanagan and the Garry Oak woodlands of southeastern Vancouver Island, the alpine meadows and glaciers of British Columbia have a fairly secure future.

The grandeur of the Tatshenshini and Muskwa-Kechika may have saved the flora and fauna of those areas from possible changes, but the forested plateaus in central and northern British Columbia will almost certainly have a very different fate. Immense areas of pine forests, from the Okanagan Highlands west to Anahim Lake and north to the Bowron River, have been cleared in attempts to stop pine beetle epidemics. The new trees planted in these clearcuts will likely never be allowed to gain the stature of the trees they replaced, since one of the strategies being used to defeat the beetles is to cut trees before they get big enough to become prime beetle habitat.

Large areas of the northern plateaus are being cut as well to produce fibre for pulp mills. Again, the new plantations produced in their place will be harvested when they are only a few decades old to maximize fibre production. Woodpeckers will find neither dead snags nor their favourite old aspens when they look for suitable sites for nesting and roosting. Songbirds that are typically found in old-growth aspen groves, such as the Connecticut Warbler, will be forced into low-quality habitats. In British Columbia, Bay-breasted, Black-throated Green and Cape May Warblers are restricted to stands of large spruce in the northeastern quarter of the province. Few if any of these stands are protected from timber harvest, and when they are gone the songbirds tied to them will be gone as well. Carnivores such as Fisher and Marten are tightly linked to the old spruce forests, where big, hollow trees provide their natural homes—they are almost never seen away from stands of older trees.

Another force altering the face of the northeast forests is that of oil and gas development. Seismic lines—long, straight lines cleared through the forest for exploration—open up huge areas to all-terrain vehicles and fragment the remaining forest into smaller and smaller squares.

These vast forests have been guarded for centuries by their own immensity, by walls of impenetrable mountain ranges, by black flies and mosquitoes and by some of the coldest winters on earth. But they are wildernesses no more, and we must strive to ensure that the tremendous changes that we bring to these areas will not forever alter the patterns and processes that ensure the survival of their natural inhabitants.

APPENDIX: MAP OF BRITISH COLUMBIA

NATURE ORGANIZATIONS

Federation of British Columbia Naturalists
1367 West Broadway
Vancouver, BC V6H 4A9
604-737-3057
This is the umbrella organization for the fifty or so naturalists' clubs in the province.

Canadian Nature Federation
1 Nicholas Street, Suite 520
Ottawa, Ontario K1N 7B7
1-800-267-4088
613-562-3447

British Columbia Field Ornithologists
P.O. Box 8059
Victoria, BC V8W 3R7

Wild Bird Trust
1489 Marine Drive, Suite 1240
West Vancouver, BC V7T 1B8
604-922-1550

FOR FURTHER READING

There are many books—ranging from the detailed and technical to the more general and accessible—that cover certain aspects of the natural history of British Columbia's mountains and northern forests. Here are a few that would be useful to the naturalist who would like to learn more.

General

Royal British Columbia Museum (formerly British Columbia Provincial Museum) Handbook series
A long series of inexpensive handbooks covering a variety of plant and animal groups, discussing their identification, distribution and ecology in British Columbia. Some are quite technical; others are not so technical. Many are out of print but are still available in libraries throughout the province.

Farley, A. L., 1979. *Atlas of British Columbia: People, environment and resource use.* Vancouver: University of British Columbia Press.

Finlay, J., and C. Finlay. 1992. *Ocean to alpine.* Edmonton: Lone Pine.

Gayton, Don. 1996. *Landscapes of the Interior.* Gabriola Island, B.C.: New Society Publishers.

Goward, T., and C. Hickson. 1995. *Nature Wells Gray.* 2d ed. Vancouver: Lone Pine.

Geology

Gabrielse, H., and C. J. Yorath, eds. 1991. *Geology of the Cordilleran orogen in Canada.* Ottawa: Geological Survey of Canada.
A thorough but technical summary of the geology of the Canadian Cordillera. A set of large colour maps is included.

Gadd, B. 1995. *Handbook of the Canadian Rockies.* 2d ed. Jasper, Alta.: Corax Press.

Harris, S. L. 1980. *Fire and ice: The Cascade volcanoes.* Rev. ed. Vancouver: Douglas & McIntyre.

Climate and Oceanography

Hare, F.K., and M.K. Thomas. 1979. *Climate Canada.* Toronto: Wiley.

Phillips, D. 1990. *The climates of Canada.* Ottawa: Environment Canada.

Ecology

Arno, S. F., and R. P. Hammerly. 1984. *Timberline: Mountain and Arctic forest frontiers.* Seattle: The Mountaineers.

Meidinger, D., and J. Pojar, eds. 1991. *Ecosystems of British Columbia.* Victoria: B.C. Ministry of Forests.
A good but somewhat technical summary of the ecological characteristics of all of British Columbia's biogeoclimatic zones.

Pielou, E.C. 1988. *The world of northern evergreens.* Ithaca: Cornell University Press.

———. 1991. *After the ice age: The return of life to glaciated North America.* Chicago: University of Chicago Press.

Botany

Douglas, G.W., G.B. Straley, and D. Meidinger, eds. 1989–1994. *The vascular plants of British Columbia.* 4 vols. Victoria: B.C. Ministry of Forests.
A thorough but technical set of volumes with identification keys and distribution summaries for every species of vascular plant in the province. No figures or maps, but still an essential set for the serious amateur botanist.

Hitchcock, C.L., and A. Cronquist. 1991. *Flora of the Pacific Northwest: An illustrated manual.* Seattle: University of Washington Press.
A compact version of the five-volume *Vascular plants of the Pacific Northwest.*

Lyons, C.P., and W. Merilees. 1995. *Trees, shrubs and flowers to know in British Columbia and Washington.* Vancouver: Lone Pine.
An updated version of Lyons's classic introduction to the common plants of the region.

MacKinnon, A., J. Pojar, and R. Coupé, eds. 1992. *Plants of northern British Columbia.* Vancouver: B.C. Ministry of Forests and Lone Pine.
Like the following book, an excellent guide to the common plants of the region.

Parish, R., R. Coupé, and D. Lloyd, eds. 1996. *Plants of southern interior British Columbia.* Vancouver: B.C. Ministry of Forests and Lone Pine.

Parish, R., and S.M. Thompson. 1994. *The tree book: Learning to recognize trees of British Columbia.* Victoria: B.C. Ministry of Forests and Canadian Forest Service.

Vitt, D.H., J.E. Marsh, and R.B. Bovey. 1988. *Mosses, lichens and ferns of northwest North America.* Edmonton: Lone Pine.

Zoology

Campbell, R.W., N.K. Dawe, I. McT. Cowan, J.M. Cooper, G.W. Kaiser, and M.C.E. McNall. 1990. *The birds of British Columbia.* Volumes 1 and 2. Victoria: Royal British Columbia Museum.

———. 1997. *The birds of British Columbia.* Volume 3. Vancouver: University of British Columbia Press.
The fourth volume is to be published within the next two years.

Corkran, C.C., and C. Thoms, eds. 1996. *Amphibians: A complete field identification manual for Oregon, Washington and British Columbia.* Vancouver: Lone Pine.

Conservation

Hammond, H. 1991. *Seeing the forest among the trees: The case for wholistic forest use.* Vancouver: Polestar.

Harding, L.E., and E. McCullum, eds. 1994. *Biodiversity in British Columbia: Our changing environment.* Delta, B.C.: Environment Canada, Canadian Wildlife Service.

INDEX

Boldface indicates photograph, figure or illustration. Species are listed by Latin name first only if this appears in the text. Otherwise, species are listed by their common names.

PHOTO AND ILLUSTRATION CREDITS

Photos

Douglas Leighton: pp. ii, x–1, 29, 37 (bottom), 47, 52–53, 65, 74–75, 82, 83, 87–88, 94, 98, 101, 104, 106

Chris Harris: pp. vi–vii

Steve Cannings: pp. 8–9, 62–63, 66, 84–85

Sydney Cannings: pp. 12–13, 33, 41, 99, 102, 103

Graham Osborne: pp. 16–17

Richard Cannings: pp. 20, 24, 25, 28, 30, 36–37, 45, 57, 70

Mark Hobson: pp. 21, 48–49, 78, 90–91, 97

David Fraser: p. 31

Robert Cannings: pp. 44, 58

Ian Jones: p. 88

Illustrations

Donald Gunn: pp. 5, 7, 26, 31, 35, 38, 40, 42, 55, 59, 73, 78

B.C. Parks: pp. 19, 27, 54, 69

Patricia Drukker-Brammall in W. B. Schofield, *Some Common Mosses of British Columbia* (Victoria: Royal British Columbia Museum, 1992): p. 59 (in box)

T. C. Brayshaw, *Trees and Shrubs of British Columbia* (Vancouver and Victoria: University of British Columbia Press and Royal British Columbia Museum, 1996): p. 60

Ralph Idema, *Manual of Nearctic Diptera*, Vol. 1, Figs 25.1, 27.1 and 31.3, based on information from Agriculture and Agri-Food, reproduced with the permission of the Minister of Supply and Services Canada, 1996: p. 61

Robert Cannings: p. 93

Canadian Forest Service: p. 71